UNITEXT

La Matematica per il 3+2

Volume 167

Editor-in-Chief

Alfio Quarteroni, Politecnico di Milano, Milan, Italy
École Polytechnique Fédérale de Lausanne (EPFL), Lausanne, Switzerland

Series Editors

Luigi Ambrosio, Scuola Normale Superiore, Pisa, Italy

Paolo Biscari, Politecnico di Milano, Milan, Italy

Ciro Ciliberto, Università di Roma "Tor Vergata", Rome, Italy

Camillo De Lellis, Institute for Advanced Study, Princeton, USA

Victor Panaretos, Institute of Mathematics, École Polytechnique Fédérale de Lausanne (EPFL), Lausanne, Switzerland

Lorenzo Rosasco, DIBRIS, Università degli Studi di Genova, Genova, Italy
Center for Brains Mind and Machines, Massachusetts Institute of Technology, Cambridge, Massachusetts, US
Istituto Italiano di Tecnologia, Genova, Italy

The **UNITEXT - La Matematica per il 3+2** series is designed for undergraduate and graduate academic courses, and also includes books addressed to PhD students in mathematics, presented at a sufficiently general and advanced level so that the student or scholar interested in a more specific theme would get the necessary background to explore it.

Originally released in Italian, the series now publishes textbooks in English addressed to students in mathematics worldwide.

Some of the most successful books in the series have evolved through several editions, adapting to the evolution of teaching curricula.

Submissions must include at least 3 sample chapters, a table of contents, and a preface outlining the aims and scope of the book, how the book fits in with the current literature, and which courses the book is suitable for.

For any further information, please contact the Editor at Springer: francesca.bonadei@springer.com

THE SERIES IS INDEXED IN SCOPUS

***

UNITEXT is glad to announce a new series of free webinars and interviews handled by the Board members, who rotate in order to interview top experts in their field.

Access this link to subscribe to the events:

https://cassyni.com/s/springer-unitext

Tommaso Ruggeri

Introduction to the Thermomechanics of Continua and Hyperbolic Systems

Tommaso Ruggeri
Dipartimento di Matematica
Università di Bologna
Bologna, Italy

ISSN 2038-5714 ISSN 2532-3318 (electronic)
UNITEXT
ISSN 2038-5722 ISSN 2038-5757 (electronic)
La Matematica per il 3+2
ISBN 978-3-031-69950-4 ISBN 978-3-031-69951-1 (eBook)
https://doi.org/10.1007/978-3-031-69951-1

© The Editor(s) (if applicable) and The Author(s), under exclusive license to Springer Nature Switzerland AG 2024

This work is subject to copyright. All rights are solely and exclusively licensed by the Publisher, whether the whole or part of the material is concerned, specifically the rights of translation, reprinting, reuse of illustrations, recitation, broadcasting, reproduction on microfilms or in any other physical way, and transmission or information storage and retrieval, electronic adaptation, computer software, or by similar or dissimilar methodology now known or hereafter developed.
The use of general descriptive names, registered names, trademarks, service marks, etc. in this publication does not imply, even in the absence of a specific statement, that such names are exempt from the relevant protective laws and regulations and therefore free for general use.
The publisher, the authors and the editors are safe to assume that the advice and information in this book are believed to be true and accurate at the date of publication. Neither the publisher nor the authors or the editors give a warranty, expressed or implied, with respect to the material contained herein or for any errors or omissions that may have been made. The publisher remains neutral with regard to jurisdictional claims in published maps and institutional affiliations.

This Springer imprint is published by the registered company Springer Nature Switzerland AG
The registered company address is: Gewerbestrasse 11, 6330 Cham, Switzerland

If disposing of this product, please recycle the paper.

To Ester, Francesca, Federico, and Sofia

Preface

The primary aim of this book is to present a unified treatment of the thermomechanics of continua using the axiomatic approach typical of rational mechanics (see, e.g., [1]). This work builds upon my previous Italian booklet designed for master's degree students in Civil Engineering at Bologna University.

Many books on continuum mechanics focus on specific types of continuous bodies, such as deformable solid bodies or fluids. The list is extensive; notable examples include the following English-language texts: [2–6]. Additionally, there are several significant Italian texts for students: [7–10].

This book, however, adopts a general perspective, presenting the mathematical structure of the balance laws and constitutive equations as a cohesive whole. Special attention is given to the modern theory of constitutive equations, particularly general principles such as the principle of material indifference and the contemporary interpretation of the principle of entropy.

I believe this book will be beneficial not only to engineering students but also to students from other scientific disciplines where chapters on continuum mechanics are studied. It provides an opportunity to view traditionally distinct topics in a broader, interconnected context.

To ensure self-consistency, the first part of the book addresses issues related to linear algebra, with a particular focus on linear operators within finite-dimensional vector spaces. This comprehensive treatment is presented in a language that directly supports the subsequent application of these concepts.

The book then offers a detailed exploration of finite deformations of continua, followed by a brief overview of kinematics. It characterizes the various forces that can exist in a continuum, introduces the stress tensor, and presents the balance laws in both Eulerian and Lagrangian forms.

Next, the modern theory of constitutive equations is defined, emphasizing the role of the general principles of material indifference and entropy as criteria for selecting physically acceptable classes of constitutive equations. The resulting field equations are specialized for various cases, including thermoelasticity, Eulerian fluids, Fourier-Navier-Stokes fluids, and rigid heat conductors.

In the final part of the book, partial differential equations in continuum mechanics are discussed, with particular attention to hyperbolic systems. The method of characteristics is introduced in both linear and non-linear cases, and the need to expand the class of solutions by introducing weak solutions is discussed, a significant example of which is represented by shock waves. As an illustrative example of weak solutions, the Riemann problem is presented for the fluid dynamic model of vehicular traffic, in which the cars are stopped at a red light and at the initial instant, the traffic light turns green.

I acknowledge that these notes, compiled into the form of a text, do not claim to be exhaustive in their coverage of this vast subject, and certain portions may not meet rigorous mathematical standards. Nevertheless, my intention in creating this text was to provide a resource for students and researchers approaching rational thermomechanics and hyperbolic differential systems for the first time.

Bologna, Italy
July 2024

Tommaso Ruggeri

Acknowledgments

I am very grateful to Takashi Arima, Francesca Brini, Andrea Mentrelli, and Shigeru Taniguchi for reviewing these notes and providing valuable suggestions.

I also want to thank Dr. Francesca Bonadies, Executive Editor at Springer, for her help in realizing this book.

Contents

Chapter 1
Matrix Operators on Vectors

Abstract In this chapter, we will present an overview of matrix operators that, when applied to a vector, transform it into another vector. The purpose is to provide the essential tools to understand the subsequent topics. We will assume familiarity with the usual operations on vectors. We denote the n-dimensional Euclidean affine space by $\mathcal{E}_n$ and its associated vector space by $\mathcal{V}$. A generic orthonormal basis of $\mathcal{V}$ will be denoted by $\{\mathbf{e}_i\}$. From now we use the Einstein convention for repeated indices, which implies summation over repeated indices. So, for instance, the scalar product between two vectors will be indicated by

$$\mathbf{a} \cdot \mathbf{b} = a_i b_i = a_1 b_1 + a_2 b_2 + \cdots + a_n b_n,$$

where a_j and b_j $(j = 1, 2, ..., n)$ are the components of vectors **a** and **b** with respect to the basis $\{\mathbf{e}_j\}$. In the case of $\mathcal{E}_3$, the vector product will be indicated by

$$\mathbf{c} = \mathbf{a} \times \mathbf{b}, \qquad c_i = \varepsilon_{ilm} a_l b_m, \quad (i = 1, 2, 3),$$

where ε_{ilm} is the Levi Civita symbol defined as

$$\varepsilon_{ilm} = \begin{cases} 0 \text{ if any two indices are equal} \\ 1 \text{ if the indices } i, l, m \text{ are an even permutation of the numbers } 1, 2, 3 \\ -1 \text{ if the indices } i, l, m \text{ are an odd permutation of the numbers } 1, 2, 3. \end{cases} \tag{1.1}$$

Therefore,

$$\begin{aligned} c_1 &= \varepsilon_{1lm} a_l b_m = \varepsilon_{123} a_2 b_3 + \varepsilon_{132} a_3 b_2 = a_2 b_3 - a_3 b_3, \\ c_2 &= \varepsilon_{2lm} a_l b_m = \varepsilon_{231} a_3 b_1 + \varepsilon_{213} a_1 b_3 = a_3 b_1 - a_1 b_3, \\ c_3 &= \varepsilon_{3lm} a_l b_m = \varepsilon_{312} a_1 b_2 + \varepsilon_{321} a_2 b_1 = a_1 b_2 - a_2 b_1. \end{aligned}$$

© The Author(s), under exclusive license to Springer Nature Switzerland AG 2024
T. Ruggeri, *Introduction to the Thermomechanics of Continua and Hyperbolic Systems*, La Matematica per il 3+2 167,
https://doi.org/10.1007/978-3-031-69951-1_1

1.1 Matrix Operators and Cartesian Components

Let us define a *matrix operator* as a linear operator $\mathbf{A}$ that transforms a generic vector $\mathbf{v} \in \mathcal{V}$ into another vector $\mathbf{w} \in \mathcal{V}$, and we will write $\mathbf{A} \in \mathcal{L}in$.[1] The linear transformation of the vector $\mathbf{v}$ into the vector $\mathbf{w}$ by the operator $\mathbf{A}$ is symbolically indicated by the expression (Fig. 1.1):

$$\mathbf{w} = \mathbf{A}\mathbf{v}.$$

Just as the vectors $\mathbf{v}$ and $\mathbf{w}$ are represented by their components v_i and w_i ($i = 1, \ldots, n$) with respect to an orthonormal basis $\{\mathbf{e}_i\}$ (*Cartesian representation*), the matrix operator $\mathbf{A}$ is represented, with respect to the same basis, by a matrix $\| A_{ij} \|$ $n \times n$ whose elements are real numbers. By definition, the components of the transformed vector $\mathbf{w}$ are calculated as follows:

$$w_i = (\mathbf{A}\mathbf{v})_i = A_{i1}v_1 + A_{i2}v_2 + \ldots + A_{in}v_n = A_{ij}v_j; \qquad i = 1, 2, \cdots n. \tag{1.2}$$

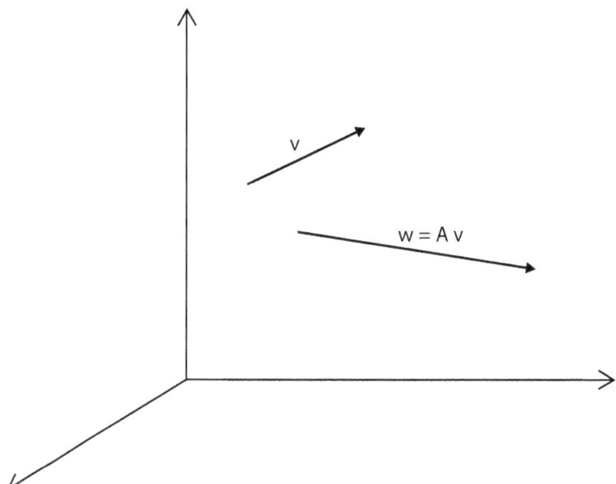

Fig. 1.1 Matrix operators

[1] Throughout the text, we will only consider linear transformations from $\mathbb{R}^n$ to $\mathbb{R}^n$ with real components.

Obviously, according to its definition as linear operator, the following properties hold:

$$\mathbf{A}(\mathbf{u}+\mathbf{v}) = \mathbf{A}\mathbf{u} + \mathbf{A}\mathbf{v} \qquad \forall \mathbf{u}, \mathbf{v} \in \mathbb{R}^n$$

$$\mathbf{A}(\lambda \mathbf{u}) = \lambda (\mathbf{A}\mathbf{u}) \qquad \forall \lambda \in \mathbb{R}, \forall \mathbf{u} \in \mathbb{R}^n.$$

Similarly to what happens for a generic vector $\mathbf{u}$, whose components u_i with respect to the orthonormal basis $\{\mathbf{e}_i\}$ depend on both the vector $\mathbf{u}$ and the basis itself according to the relation

$$u_i = \mathbf{u} \cdot \mathbf{e}_i, \tag{1.3}$$

the elements A_{ij} of the matrix representing the operator $\mathbf{A}$ depend on both the operator itself and the basis. In fact, we have

$$A_{ij} = \mathbf{e}_i \cdot \mathbf{A}\mathbf{e}_j, \tag{1.4}$$

which can be easily verified by using (1.2), taking into account that $\mathbf{e}_1 \equiv (1, 0, \cdots, 0)$, $\mathbf{e}_2 \equiv (0, 1, \cdots, 0)$, $\cdots$, $\mathbf{e}_n \equiv (0, 0, \cdots, 1)$:

$$\mathbf{A}\mathbf{e}_1 \equiv (A_{11}, A_{21}, \ldots, A_{n1})$$

$$\mathbf{A}\mathbf{e}_2 \equiv (A_{12}, A_{22}, \ldots, A_{n2})$$

$$\ldots$$

$$\mathbf{A}\mathbf{e}_n \equiv (A_{1n}, A_{2n}, \ldots, A_{nn}),$$

which allows us to write, for $j = 1, 2, \ldots, n$,

$$\mathbf{A}\mathbf{e}_j \equiv \left(A_{1j}, A_{2j}, \ldots, A_{nj}\right). \tag{1.5}$$

Therefore, the components of the vector $\mathbf{A}\mathbf{e}_j$ are the j-th column of the matrix associated with the operator $\mathbf{A}$, and the validity of (1.4) can be easily verified.

In the following, we will often refer to spaces of dimension $n = 3$, in which matrix operators $\mathbf{A}$ are represented, with respect to an assigned basis, by 3×3 matrices.

Remark 1 It is essential to note that the vector $\mathbf{w}$ depends solely on the vector $\mathbf{v}$ and the operator $\mathbf{A}$, being independent of the basis used to represent $\mathbf{v}$, $\mathbf{w}$, and $\mathbf{A}$. Therefore, we must establish an appropriate set of transformation rules for vector components and operators, a topic that we will explore in Sect. 1.16. At times, the components of vectors are referred to as *rank-1 tensors*, while the components of matrix operators are denoted as *rank-2 tensors* or *double tensors* (or simply *tensors*). It is worth mentioning that the formal concept of a tensor is more extensive and intricate, but its detailed exploration falls beyond the scope of present text.

1.2 Identity Operator

The *identity operator*, denoted by **I**, is a linear operator that transforms any vector **v** into itself:

$$\mathbf{v} = \mathbf{I}\mathbf{v} \quad \forall \mathbf{v} \in \mathbb{R}^n .$$

It can be easily verified that the components of the identity operator **I** with respect to any given basis are given by the KRONECKER δ_{ij} symbol:

$$\mathbf{I} \equiv \left\| \delta_{ij} \right\| , \qquad \delta_{ij} = \begin{cases} 0 & \text{for } i \neq j \\ 1 & \text{for } i = j \, . \end{cases}$$

1.3 Scalar Multiplication of a Matrix Operator

The scalar multiplication of an operator **A** by a scalar $\lambda \in \mathbb{R}$ is defined as the operator **C** given by

$$\mathbf{C} = \lambda \mathbf{A},$$

which, for any vector **v** satisfying $\mathbf{A}\mathbf{v} = \mathbf{w}$, yields

$$\mathbf{C}\mathbf{v} = \lambda \mathbf{w}.$$

If we represent the matrix operators **A** and **C** with respect to a chosen basis, the generic component C_{ij} of **C** is related to the corresponding component A_{ij} of **A** by the equation:

$$C_{ij} = \lambda A_{ij} .$$

An operator of the form

$$\mathbf{A} = \lambda \mathbf{I}, \tag{1.6}$$

where **I** is the identity operator, is called an *isotropic operator* for reasons that will be discussed later.

1.4 Sum of Two Operators

Given two operators **A** and **B**, the sum of the operators, denoted by **C**, is defined as the operator that satisfies

$$\mathbf{C}\mathbf{v} = \mathbf{A}\mathbf{v} + \mathbf{B}\mathbf{v} \quad \forall \mathbf{v},$$

and it is written as

$$\mathbf{C} = \mathbf{A} + \mathbf{B}.$$

It can be easily verified that such an operator exists, and if we represent the operators **A**, **B**, and **C** with respect to a chosen basis, the generic component C_{ij} of **C** is related to the corresponding components A_{ij} and B_{ij} of **A** and **B**, respectively, by the equation:

$$C_{ij} = A_{ij} + B_{ij}.$$

1.5 Product of Two Operators

We can also define the product **C** of two operators **A** and **B** ($\mathbf{A}, \mathbf{B}, \mathbf{C} \in \mathcal{L}in$) as follows:

$$\mathbf{C}\mathbf{v} = \mathbf{A}(\mathbf{B}\mathbf{v}) \quad \forall \mathbf{v} \in \mathbb{R}^n$$

and denote it by $\mathbf{C} = \mathbf{A}\mathbf{B}$. Given a basis, the element C_{ij} of the matrix representing the operator **C** is given by

$$C_{ij} = A_{ik} B_{kj},$$

where $\|A_{ij}\|$ and $\|B_{ij}\|$ represent the matrices associated with the operators **A** and **B**, respectively, in the same basis used to represent $\|C_{ij}\|$.

It is important to note that the product of operators does not generally obey the commutative property, i.e.,

$$\mathbf{A}\mathbf{B} \neq \mathbf{B}\mathbf{A}.$$

If $\mathbf{A}\mathbf{B} = \mathbf{B}\mathbf{A}$ holds, it is said that the two operators **A** and **B** commute with each other.

It can also be easily verified that any operator $\mathbf{A} \in \mathcal{L}in$ commutes with the identity operator **I**, i.e.,

$$\mathbf{A}\mathbf{I} = \mathbf{I}\mathbf{A}.$$

1.6 Transpose Operator

The *transpose operator* of an operator $\mathbf{A}$, denoted by $\mathbf{A}^T$, is the linear operator that satisfies the following relation:

$$\mathbf{A}\mathbf{v} \cdot \mathbf{w} = \mathbf{v} \cdot \mathbf{A}^T\mathbf{w}, \quad \forall \mathbf{v}, \mathbf{w}. \tag{1.7}$$

In a chosen basis, the matrix associated with the transpose operator $\mathbf{A}^T$ can be obtained by interchanging the rows and columns of the matrix associated with the operator $\mathbf{A}$. Specifically,

$$\begin{aligned}
&\mathbf{A}\mathbf{v} \cdot \mathbf{w} = \mathbf{v} \cdot \mathbf{A}^T\mathbf{w}, \\
&(\mathbf{A}\mathbf{v})_j\, w_j = v_i \left(\mathbf{A}^T\mathbf{w}\right)_i, \\
&A_{ji} v_i w_j = v_i A^T_{ij} w_j,
\end{aligned}$$

which implies

$$A^T_{ij} = A_{ji}.$$

1.7 Trace of an Operator

The *trace* of an operator $\mathbf{A}$, denoted by $\operatorname{tr}\mathbf{A}$, is the scalar obtained by summing the elements on the main diagonal of the matrix associated with the operator $\mathbf{A}$:

$$\operatorname{tr}\mathbf{A} = A_{ii}.$$

The trace is a linear operator, and it satisfies the following properties:

- $\operatorname{tr}(\mathbf{A} + \mathbf{B}) = \operatorname{tr}\mathbf{A} + \operatorname{tr}\mathbf{B}$ for all $\mathbf{A}, \mathbf{B} \in \mathcal{L}in$.
- $\operatorname{tr}(\lambda\mathbf{A}) = \lambda \operatorname{tr}\mathbf{A}$ for all $\lambda \in \mathbb{R}$ and $\mathbf{A} \in \mathcal{L}in$.

An operator $\mathbf{A}$ with zero trace is called *deviatoric*. Some immediate identities involving the trace are:

- $\operatorname{tr}\mathbf{I} = n$, where n is the dimension of the vector space.
- $\operatorname{tr}\mathbf{A} = \operatorname{tr}\left(\mathbf{A}^T\right)$.
- $\operatorname{tr}(\mathbf{A}\mathbf{B}) = \operatorname{tr}(\mathbf{B}\mathbf{A})$.

The first two identities are straightforward, while the last one can be shown as follows:

$$\operatorname{tr}(\mathbf{A}\mathbf{B}) = (\mathbf{A}\mathbf{B})_{ii} = A_{ij}B_{ji} = (\mathbf{B}\mathbf{A})_{jj} = \operatorname{tr}(\mathbf{B}\mathbf{A}).$$

As a consequence of this property, we have

$$\mathrm{tr}(\mathbf{ABC}) = \mathrm{tr}(\mathbf{BCA}) = \mathrm{tr}(\mathbf{CAB}), \tag{1.8}$$

which tells us that we can cyclically permute the operators **A**, **B**, and **C** without changing the trace.

In Sect. 1.16.1, we will demonstrate that the trace is independent of the chosen basis.

1.8 Determinant of an Operator

We define the *determinant* of an operator **A** of order n with associated matrix $\|A_{ij}\|$, denoted by $\det \mathbf{A}$, as the scalar calculated as follows:

- If $n = 1$,

$$\det \mathbf{A} = A_{11}.$$

- If $n > 1$ and we fix an i $(1 \leq i \leq n)$,

$$\det \mathbf{A} = \sum_{j=1}^{n} (-1)^{i+j} A_{ij} \det \mathbf{A}^{(ij)},$$

  where $\mathbf{A}^{(ij)}$ is the operator of order $(n-1)$ whose associated matrix is obtained by removing the i-th row and j-th column from the matrix associated with **A**.

Note that the definition is iterative, as the determinant of an operator of order n is defined as a linear combination of n determinants of operators of order $(n-1)$. We will prove in Sect. 1.16.1 that the determinant does not depend on the choice of the basis. Starting from the definition of the determinant of an operator $\mathbf{A} \in \mathcal{L}in$, it can be verified that the following important properties hold:

$$\det \mathbf{A} = \det \mathbf{A}^T, \tag{1.9}$$

$$\det(\lambda \mathbf{A}) = \lambda^n \det \mathbf{A} \quad (\lambda \in \mathbb{R}), \tag{1.10}$$

$$\det(\mathbf{AB}) = \det \mathbf{A} \det \mathbf{B}. \tag{1.11}$$

The last property is known as *Binet's theorem.*

To indicate that a non-singular operator **A** has a positive determinant, we write $\mathbf{A} \in \mathcal{L}in^+$.

1.8.1 Expression of the Determinant for n = 3

In the case of spaces with dimension $n = 3$, which are of particular interest due to the definition of the cross product, we can verify, based on the well-known properties of the triple product and Eq. (1.5), that the following relationship holds:

$$\det \mathbf{A} = \mathbf{A}\mathbf{e}_1 \times \mathbf{A}\mathbf{e}_2 \cdot \mathbf{A}\mathbf{e}_3 \tag{1.12}$$

and also the more general form:

$$\mathbf{e}_i \times \mathbf{e}_j \cdot \mathbf{e}_k \ \det \mathbf{A} = \mathbf{A}\mathbf{e}_i \times \mathbf{A}\mathbf{e}_j \cdot \mathbf{A}\mathbf{e}_k \quad \forall i, j, k = 1, 2, 3, \tag{1.13}$$

where $\{\mathbf{e}_i\}$ is an orthonormal basis.

1.9 Inverse Operator

An operator $\mathbf{A}$ is said to be invertible if there exists an operator $\mathbf{B}$ such that

$$\mathbf{B}\mathbf{A} = \mathbf{A}\mathbf{B} = \mathbf{I}. \tag{1.14}$$

If $\mathbf{B}$ exists, it is unique and is called the *inverse* of $\mathbf{A}$, denoted by $\mathbf{A}^{-1}$.

Note that this operator (if it exists) allows us to perform the inverse transformation of the one carried out by the operator $\mathbf{A}$, that is (Fig. 1.2):

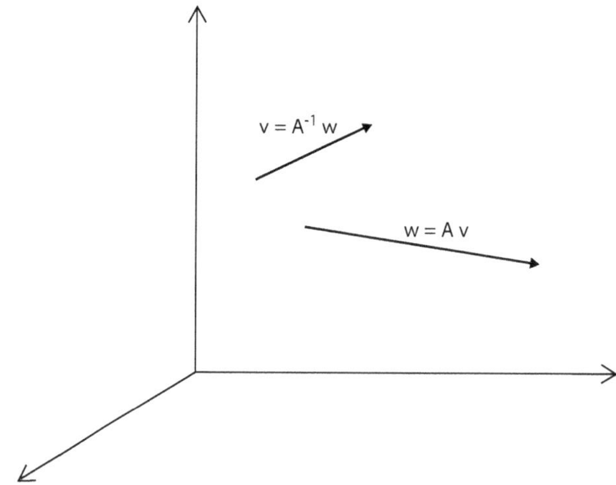

Fig. 1.2 Inverse operator

$$\text{If} \quad \mathbf{w} = \mathbf{A}\mathbf{v} \qquad \text{then} \ \ \mathbf{v} = \mathbf{A}^{-1}\mathbf{w} \quad \forall \mathbf{v} \in \mathbb{R}^n. \tag{1.15}$$

The uniqueness of the operator, if it exists, can be easily verified. If there were another operator $\widehat{\mathbf{B}}$ satisfying (1.14), then it would also satisfy

$$\widehat{\mathbf{B}}\mathbf{A} = \mathbf{I}. \tag{1.16}$$

Multiplying this expression from the right by $\mathbf{B}$ and considering the second identity of (1.14), we immediately obtain $\widehat{\mathbf{B}} = \mathbf{B}$. In order to prove the existence of $\mathbf{B}$, we first need to define the cofactor operator.

1.10 Cofactor Operator

Let $\mathbf{A}$ be a linear operator $n \times n$ with $n > 1$. We define the *cofactor operator* of $\mathbf{A}$, denoted by $\mathbf{A}^C$, as the operator whose components, with respect to a given basis, are the corresponding cofactor of $\mathbf{A}$:

$$A^C_{ij} = (-1)^{i+j} \det \mathbf{A}^{(ij)}, \tag{1.17}$$

where $\mathbf{A}^{(ij)}$ is the matrix obtained from the matrix $\mathbf{A}$ by deleting the i-th row and j-th column. Now let us recall the two LAPLACE theorems on determinants:

First Laplace Theorem The sum of the products of the elements of a row or a column of a matrix by their respective cofactors is equal to the determinant of the matrix.

Second Laplace Theorem The sum of the products of the elements of a row or column of a matrix by the cofactors of another row or column of the matrix is equal to zero.

These two theorems imply the following equations:

$$A_{ik} A^C_{ij} = \det \mathbf{A} \ \delta_{jk} \qquad \text{or} \qquad A_{ij} A^C_{kj} = \det \mathbf{A} \ \delta_{ik}, \tag{1.18}$$

where δ_{jk} is the Kronecker symbol.

In terms of operators, these equations can be written as

$$\mathbf{A}^{CT}\mathbf{A} = \det \mathbf{A}\, \mathbf{I} \qquad \text{or} \qquad \mathbf{A}\mathbf{A}^{CT} = \det \mathbf{A}\, \mathbf{I}. \tag{1.19}$$

Now we can easily prove the following theorem:

Theorem 1.10.1 *An operator* $\mathbf{A}$ *is invertible if and only if* $\det \mathbf{A} \neq 0$. *In this case, we have*

$$\mathbf{A}^{-1} = \frac{\mathbf{A}^{CT}}{\det \mathbf{A}} \tag{1.20}$$

or in terms of components

$$A_{ij}^{-1} = \frac{A_{ji}^{C}}{\det \mathbf{A}}.$$

Proof If the inverse operator exists, from (1.14) and the theorem of BINET (1.11), we have

$$\det \mathbf{B}\ \det \mathbf{A} = 1$$

and thus necessarily $\det \mathbf{A} \neq 0$. Vice versa, if $\det \mathbf{A} \neq 0$, from (1.19) we have

$$\left(\frac{1}{\det \mathbf{A}}\mathbf{A}^{CT}\right)\mathbf{A} = \mathbf{A}\left(\frac{1}{\det \mathbf{A}}\mathbf{A}^{CT}\right) = \mathbf{I}.$$

Comparing this with (1.14), we immediately see that $\mathbf{A}$ is invertible and that (1.20) holds.

It can also be shown that if $\det \mathbf{A} \neq 0$, then $\mathbf{A}$ is injective and surjective, and we can write

$$\mathbf{A}^{-1}\mathbf{A} = \mathbf{A}\mathbf{A}^{-1} = \mathbf{I}. \tag{1.21}$$

□

Remark 2 We defined the cofactor operator using the expression (1.17), which excludes the case where the operator reduces to a scalar, i.e., when $n = 1$. However, it is useful to define the cofactor operator even in the case of $n = 1$. From (1.20), we have

$$\mathbf{A}^{C} = \mathbf{A}^{-1T} \det \mathbf{A}. \tag{1.22}$$

Since the right-hand side of (1.22) is well defined for the limiting case of $n = 1$ and equals 1, we take the definition that the operator $\mathbf{A}^C$ equals 1 when $n = 1$, and for any $n > 1$, it is defined by (1.17).

1.11 Some Notable Identities of Matrix Operators

Let us state some notable identities:

$$(\mathbf{AB})^T = \mathbf{B}^T \mathbf{A}^T \tag{1.23}$$

$$(\mathbf{AB})^{-1} = \mathbf{B}^{-1} \mathbf{A}^{-1} \qquad \text{(for non-singular } \mathbf{A} \text{ and } \mathbf{B}) \tag{1.24}$$

$$(\mathbf{AB})^C = \mathbf{A}^C \mathbf{B}^C \tag{1.25}$$

$$\left(\mathbf{A}^{-1}\right)^T = \left(\mathbf{A}^T\right)^{-1} \qquad \text{(for non-singular } \mathbf{A}) \tag{1.26}$$

$$\left(\mathbf{A}^C\right)^T = \left(\mathbf{A}^T\right)^C \tag{1.27}$$

$$\left(\mathbf{A}^C\right)^C = (\det \mathbf{A})^{n-2} \mathbf{A}, \qquad \forall n \in \mathbb{N} \geq 2. \tag{1.28}$$

- *Proof of identity* (1.23): Using the property (1.7), for two vectors $\mathbf{v}$ and $\mathbf{w}$, we have

$$(\mathbf{AB})\, \mathbf{v} \cdot \mathbf{w} = \mathbf{v} \cdot (\mathbf{AB})^T\, \mathbf{w}$$

  and also (see Sect. 1.5)

$$(\mathbf{AB})\, \mathbf{v} \cdot \mathbf{w} = \mathbf{A}\, (\mathbf{Bv}) \cdot \mathbf{w} = \mathbf{Bv} \cdot \mathbf{A}^T \mathbf{w} = \mathbf{v} \cdot \mathbf{B}^T \mathbf{A}^T \mathbf{w}.$$

  Comparing the two equations, we conclude that $(\mathbf{AB})^T = \mathbf{B}^T \mathbf{A}^T$.
- *Proof of identity* (1.24): Let $\mathbf{v}$ and $\mathbf{w}$ be two vectors such that

$$\mathbf{w} = (\mathbf{AB})\, \mathbf{v} = \mathbf{A}\, (\mathbf{Bv})\,.$$

  From the first equation (assuming $\mathbf{A}$ and $\mathbf{B}$ are non-singular), we have

$$\mathbf{v} = (\mathbf{AB})^{-1}\, \mathbf{w}, \tag{1.29}$$

  and from the second equation, we have

$$\mathbf{Bv} = \mathbf{A}^{-1} \mathbf{w}.$$

  Multiplying the latter equation on the left by the operator $\mathbf{B}^{-1}$, we obtain

$$\mathbf{B}^{-1}\, (\mathbf{Bv}) = \mathbf{v} = \mathbf{B}^{-1} \left(\mathbf{A}^{-1} \mathbf{w}\right) = \mathbf{B}^{-1} \mathbf{A}^{-1} \mathbf{w}. \tag{1.30}$$

By comparing equation (1.29) with (1.30), we obtain identity (1.24). Taking $\mathbf{B} = \lambda\mathbf{I}$ yields a particular case of (1.24):

$$(\lambda\mathbf{A})^{-1} = \frac{1}{\lambda}\mathbf{A}^{-1}. \tag{1.31}$$

- *Proof of identity* (1.25): From (1.17), we have

$$(\mathbf{AB})^C_{ij} = (-1)^{i+j}\det(\mathbf{AB})^{(ij)} \tag{1.32}$$

and at the same time

$$\begin{aligned}(\mathbf{A}^C\mathbf{B}^C)_{ij} &= A^C_{ik}\,B^C_{kj}\\ &= (-1)^{i+k}\det\mathbf{A}^{(ik)}(-1)^{k+j}\det\mathbf{B}^{(kj)}\\ &= (-1)^{i+j+2k}\det\mathbf{A}^{(ik)}\det\mathbf{B}^{(kj)}\\ &= (-1)^{i+j}\det\left(\mathbf{A}^{(ik)}\mathbf{B}^{(kj)}\right),\end{aligned}$$

where in the last step we used the fact that $(-1)^{2k} = 1$ for all $k \in \mathbb{N}$ and applied BINET'S theorem. At this point, it is not difficult to see that

$$\mathbf{A}^{(ik)}\mathbf{B}^{(kj)} = (\mathbf{AB})^{(ij)}, \tag{1.33}$$

and thus we can write

$$(\mathbf{A}^C\mathbf{B}^C)_{ij} = (-1)^{i+j}\det(\mathbf{AB})^{(ij)}. \tag{1.34}$$

By comparing equation (1.32) with (1.34), we can conclude that the identity (1.25) holds.
- *Proof of identity* (1.26): From the equation

$$\mathbf{A}^T\left(\mathbf{A}^{-1}\right)^T = \left(\mathbf{A}^{-1}\mathbf{A}\right)^T = \mathbf{I},$$

we can immediately obtain, by multiplying the first and last terms on the left by $(\mathbf{A}^T)^{-1}$,

$$\left(\mathbf{A}^{-1}\right)^T = \left(\mathbf{A}^T\right)^{-1}, \tag{1.35}$$

which proves the statement.

- *Proof of identity* (1.27): Using the definition of $\mathbf{A}^{(ij)}$, we have the identity

$$\left(\mathbf{A}^{(ij)}\right)^T = \left(\mathbf{A}^T\right)^{(ji)}. \tag{1.36}$$

Taking into account (1.17), we have

$$\left(\mathbf{A}^{CT}\right)_{ij} = A^C_{ji} = (-1)^{j+i} \det \mathbf{A}^{(ji)}, \tag{1.37}$$

while using (1.17), (1.36), and (1.9), we obtain

$$\begin{aligned}\left(\mathbf{A}^{TC}\right)_{ij} &= (-1)^{i+j} \det\left(\left(\mathbf{A}^T\right)^{(ij)}\right) = (-1)^{i+j} \det\left(\left(\mathbf{A}^{(ji)}\right)^T\right)\\ &= (-1)^{i+j} \det \mathbf{A}^{(ji)}.\end{aligned} \tag{1.38}$$

By comparing (1.38) with (1.37), we can conclude that the identity (1.27) holds.
An alternative and simpler proof can be given in the case where $\mathbf{A}$ is non-singular. From (1.20), we have

$$\left(\mathbf{A}^C\right)^T = \det \mathbf{A}\, \mathbf{A}^{-1}.$$

On the other hand, using (1.20) and (1.35), we also have

$$\left(\mathbf{A}^T\right)^C = \det \mathbf{A} \left(\mathbf{A}^{-1T}\right)^T = \det \mathbf{A}\, \mathbf{A}^{-1} = \left(\mathbf{A}^C\right)^T.$$

This proves the statement.
- *Proof of identity* (1.28): We provide the proof in the case where $\mathbf{A}$ is non-singular, but the property holds for singular matrices as well. From (1.20), we have

$$\mathbf{B} = \mathbf{A}^C = (\det \mathbf{A})\, \mathbf{A}^{-1T} \quad \text{and hence} \quad \left(\mathbf{A}^C\right)^C = \mathbf{B}^C = (\det \mathbf{B})\, \mathbf{B}^{-1T}. \tag{1.39}$$

By using $(1.39)_1$ and (1.10), we have

$$\det \mathbf{B} = (\det \mathbf{A})^{n-1}, \quad \text{and} \quad \mathbf{B}^{-1T} = \left(\mathbf{B}^T\right)^{-1} = \frac{1}{\det \mathbf{A}} \mathbf{A}. \tag{1.40}$$

Substituting (1.40) into $(1.39)_2$ proves the identity (1.28). Note that, except for the case $n = 2$, if $\mathbf{A}$ is singular, we have $\left(\mathbf{A}^C\right)^C = 0$ for all $n > 2$.

1.11.1 Some Notable Identities for $n = 3$

For vector spaces of dimension $n = 3$, we can state the following identities of significant importance for applications:

$$\mathbf{Au} \times \mathbf{Av} \cdot \mathbf{Aw} = (\det \mathbf{A})\, \mathbf{u} \times \mathbf{v} \cdot \mathbf{w} \tag{1.41}$$

$$\mathbf{Au} \times \mathbf{Av} = \mathbf{A}^C (\mathbf{u} \times \mathbf{v})\,. \tag{1.42}$$

- *Proof of identity* (1.41): Using (1.13), we can write

$$\begin{aligned}\mathbf{Au} \times \mathbf{Av} \cdot \mathbf{Aw} &= \mathbf{A}(u_i\mathbf{e}_i) \times \mathbf{A}\left(v_j\mathbf{e}_j\right) \cdot \mathbf{A}(w_k\mathbf{e}_k) \\ &= u_i v_j w_k\, \mathbf{Ae}_i \times \mathbf{Ae}_j \cdot \mathbf{Ae}_k \\ &= u_i v_j w_k\, (\det \mathbf{A})\, \mathbf{e}_i \times \mathbf{e}_j \cdot \mathbf{e}_k \\ &= \det \mathbf{A}\, (u_i\mathbf{e}_i) \times \left(v_j\mathbf{e}_j\right) \cdot (w_k\mathbf{e}_k) \\ &= (\det \mathbf{A})\, \mathbf{u} \times \mathbf{v} \cdot \mathbf{w},\end{aligned}$$

  and the identity is proved.
- *Proof of identity* (1.42): Taking a generic nonzero vector $\mathbf{w} \in \mathbb{R}^3$, we can write

$$\mathbf{Au} \times \mathbf{Av} \cdot \mathbf{Aw} = \mathbf{A}^T (\mathbf{Au} \times \mathbf{Av}) \cdot \mathbf{w},$$

  but according to (1.41), we also have

$$\mathbf{Au} \times \mathbf{Av} \cdot \mathbf{Aw} = (\det \mathbf{A})\, \mathbf{u} \times \mathbf{v} \cdot \mathbf{w}.$$

  Equating the right-hand sides of the two previous identities, we get

$$\mathbf{A}^T (\mathbf{Au} \times \mathbf{Av}) = (\det \mathbf{A})\, \mathbf{u} \times \mathbf{v},$$

  and therefore,

$$\mathbf{Au} \times \mathbf{Av} = (\det \mathbf{A})\, \mathbf{A}^{-1T} (\mathbf{u} \times \mathbf{v}) = \mathbf{A}^C (\mathbf{u} \times \mathbf{v})\,.$$

1.12 Scalar Product Between Operators

We define the *scalar product* between two operators $\mathbf{A}$ and $\mathbf{B}$ as the scalar:

$$\mathbf{A} \cdot \mathbf{B} = \operatorname{tr}\left(\mathbf{AB}^T\right).$$

It is worth noting that the scalar product between operators has the following properties:

- Commutativity: $\mathbf{A} \cdot \mathbf{B} = \mathbf{B} \cdot \mathbf{A}$.
- Distributive property with respect to addition: $\mathbf{A} \cdot (\mathbf{B} + \mathbf{C}) = \mathbf{A} \cdot \mathbf{B} + \mathbf{A} \cdot \mathbf{C}$.

The commutativity of the scalar product between operators can be demonstrated by observing that

$$\mathbf{A} \cdot \mathbf{B} = \operatorname{tr}\left(\mathbf{A}\mathbf{B}^T\right) = \operatorname{tr}\left(\left(\mathbf{A}\mathbf{B}^T\right)^T\right) = \operatorname{tr}\left(\mathbf{B}\mathbf{A}^T\right) = \mathbf{B} \cdot \mathbf{A}.$$

The distributive property can be demonstrated by noting that

$$\begin{aligned}\mathbf{A} \cdot (\mathbf{B} + \mathbf{C}) &= \operatorname{tr}\left(\mathbf{A}\left(\mathbf{B} + \mathbf{C}\right)^T\right) = \operatorname{tr}\left(\mathbf{A}\mathbf{B}^T + \mathbf{A}\mathbf{C}^T\right) \\ &= \operatorname{tr}\left(\mathbf{A}\mathbf{B}^T\right) + \operatorname{tr}\left(\mathbf{A}\mathbf{C}^T\right) = \mathbf{A} \cdot \mathbf{B} + \mathbf{A} \cdot \mathbf{C}.\end{aligned}$$

It is important to remind that, since the trace of an operator is invariant under a change of basis, the scalar product between operators is also invariant.

Given the components A_{ij} and B_{ij} of the operators $\mathbf{A}$ and $\mathbf{B}$ with respect to a basis, the scalar product can be written as

$$\mathbf{A} \cdot \mathbf{B} = \operatorname{tr}\left(\mathbf{A}\mathbf{B}^T\right) = \left(\mathbf{A}\mathbf{B}^T\right)_{ii} = A_{ij} B^T_{ji} = A_{ij} B_{ij}.$$

In particular, we have

$$\mathbf{A} \cdot \mathbf{A} = A_{ij} A_{ij} \geq 0 \quad \forall \mathbf{A} \in \mathcal{L}in \quad (\mathbf{A} \cdot \mathbf{A} = 0 \quad \Leftrightarrow \quad \mathbf{A} = \mathbf{0})\,. \tag{1.43}$$

This suggests using the scalar product to define a possible norm in the space of matrix operators:

$$\|\mathbf{A}\| = \sqrt{\mathbf{A} \cdot \mathbf{A}}.$$

This norm is known as FROBENIUS NORM (*or Euclidean norm*).[2] Two nonzero operators such that $\mathbf{A} \cdot \mathbf{B} = 0$ are said to be *orthogonal* to each other.

[2] It can be easily verified that the norm defined in this way satisfies, in addition to (1.43), the other necessary properties:

$$\|\mathbf{A} + \mathbf{B}\| \leq \|\mathbf{A}\| + \|\mathbf{B}\| \quad \forall \mathbf{A}, \mathbf{B} \in \mathcal{L}in$$

$$\|\alpha \mathbf{A}\| = |\alpha| \, \|\mathbf{A}\| \quad \forall \alpha \in \mathbb{R}, \ \forall \mathbf{A} \in \mathcal{L}in.$$

1.13 Symmetric and Antisymmetric Operators

An operator $\mathbf{S}$ is said to be *symmetric* if it satisfies the condition:

$$\mathbf{S} = \mathbf{S}^T.$$

For the components of the matrix representing a symmetric operator, regardless of the basis used for its representation, the following relation holds:

$$S_{ij} = S_{ji}.$$

We use the notation $\mathbf{S} \in \mathcal{Sym}$ to indicate that an operator $\mathbf{S}$ is symmetric.

An operator $\mathbf{A}$ is said to be *antisymmetric* or (*skew-symmetric*) if it satisfies the condition:

$$\mathbf{A} = -\mathbf{A}^T.$$

For the components of the matrix representing an antisymmetric operator, regardless of the basis used for its representation, the following relation holds:

$$A_{ij} = -A_{ji}.$$

We use the notation $\mathbf{A} \in \mathcal{Skw}$ to indicate that an operator $\mathbf{A}$ is antisymmetric.

We define the *significant components* of an operator $\mathbf{A}$ as the minimum number of elements that need to be assigned in order to uniquely determine the matrix representing the operator in a given basis.

For a general operator of order $n = 3$, the significant components are 9, which is equal to the number of components of a matrix $\| F_{ij} \|$:

$$\mathbf{F} \equiv \begin{pmatrix} \times & \times & \times \\ \times & \times & \times \\ \times & \times & \times \end{pmatrix}.$$

For a symmetric matrix $\mathbf{S} \in \mathcal{Sym}$, the significant components are 6, since the relation $S_{ij} = S_{ji}$ holds:

$$\mathbf{S} \equiv \begin{pmatrix} \times & \times & \times \\ & \times & \times \\ & & \times \end{pmatrix}.$$

For an antisymmetric matrix $\mathbf{A} \in \mathcal{Skw}$, the significant components are only 3. Notice that, since $A_{ij} = -A_{ji}$, the elements on the main diagonal must be zero:

$$\mathbf{A} \equiv \begin{pmatrix} 0 & \times & \times \\ & 0 & \times \\ & & 0 \end{pmatrix}.$$

Thus, it is necessary and sufficient to assign three components to uniquely determine an antisymmetric operator $\mathbf{A}$.

The above statements can be generalized to spaces of any dimension n as follows:

- The significant components of a general operator are equal to

$$n^2.$$

- The significant components of a symmetric operator are equal to

$$\frac{n^2}{2} + \frac{n}{2} = \frac{n}{2}(n+1).$$

- The significant components of an antisymmetric operator are equal to

$$\frac{n^2}{2} - \frac{n}{2} = \frac{n}{2}(n-1).$$

Finally, we mention two interesting properties of antisymmetric operators:

1. Every antisymmetric operator transforms any vector $\mathbf{v}$ into a vector orthogonal to $\mathbf{v}$. The proof of this property is straightforward:

$$\mathbf{Av} \cdot \mathbf{v} = \mathbf{v} \cdot \mathbf{A}^T \mathbf{v} = -\mathbf{v} \cdot \mathbf{Av} \quad \Rightarrow \quad \mathbf{Av} \cdot \mathbf{v} = 0 \tag{1.44}$$

   (see Fig. 1.3).
2. Every antisymmetric operator in spaces of odd dimension n is singular. Indeed, using (1.10), for $\mathbf{A} \in \mathcal{Skw}$, we have

$$\det \mathbf{A} = \det \mathbf{A}^T = \det(-\mathbf{A}) = (-1)^n \det \mathbf{A} \quad \Rightarrow \quad \det \mathbf{A} = 0 \ \ \forall\, n \text{ odd}.$$

   This implies that *antisymmetric operators in spaces of odd dimension (including $\mathbb{R}^3$) do not have an inverse operator.*

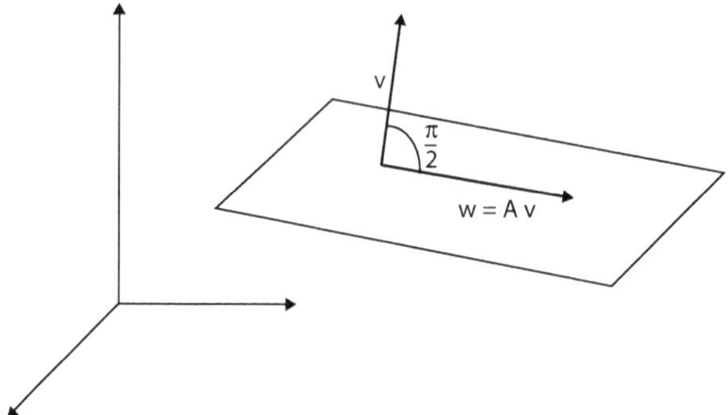

Fig. 1.3 Operators antisymmetric

1.13.1 Dual Vector Associated with an Antisymmetric Operator

In a three-dimensional space ($n = 3$), an antisymmetric operator $\mathbf{A}$ has three significant components (e.g., A_{12}, A_{13}, and A_{23}). It is possible to associate a vector $\boldsymbol{\omega}$ with it, whose components in the same basis as the components of $\mathbf{A}$ are defined as follows:

$$\omega_1 = A_{32} = -A_{23},$$
$$\omega_2 = A_{13} = -A_{31},$$
$$\omega_3 = A_{21} = -A_{12}.$$

Therefore, we can write

$$\mathbf{A} = \begin{pmatrix} 0 & -\omega_3 & \omega_2 \\ \omega_3 & 0 & -\omega_1 \\ -\omega_2 & \omega_1 & 0 \end{pmatrix},$$

and it is easy to verify that for any vector $\mathbf{v}$, the relation

$$\mathbf{A}\mathbf{v} = \boldsymbol{\omega} \times \mathbf{v} \tag{1.45}$$

holds. This equation expresses the fact that the application of the antisymmetric operator $\mathbf{A}$ to an arbitrary vector $\mathbf{v}$ is equivalent to left vector multiplying $\mathbf{v}$ by the *dual vector* $\boldsymbol{\omega}$ associated with the operator $\mathbf{A}$ (see Fig. 1.4). Note that this is consistent with the previous observation that $\mathbf{A}\mathbf{v}$ is always orthogonal to $\mathbf{v}$. It is worth noting that in spaces of dimension $n \neq 3$, the number of significant components of an antisymmetric matrix always differs from the number of components of a vector. Specifically, excluding the nonsignificant case $n = 0$, we have $n(n-1)/2 = n$ if

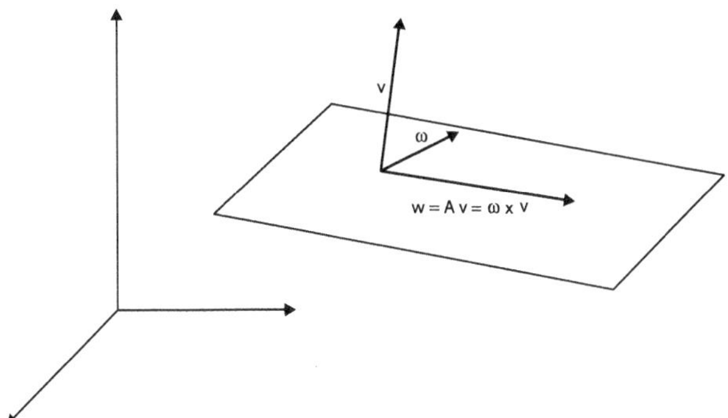

Fig. 1.4 Dual vector

and only if $n = 3$. This explains why it is not possible to define a dual vector and, therefore, an associated cross product in dimensions other than 3.

It should be observed that the operation of $\mathbf{Av}$ is always defined, and therefore, if desired, the concept of cross product can be extended to n-dimensional spaces in this sense.

Theorem 1.13.1 *A symmetric operator* $\mathbf{S}$ *and an antisymmetric operator* $\mathbf{A}$ *are always orthogonal to each other, that is,*

$$\textit{If} \quad \mathbf{S} \in \mathcal{S}ym, \textit{ and } \mathbf{A} \in \mathcal{S}kw \ \Rightarrow \ \mathbf{S} \cdot \mathbf{A} = 0. \tag{1.46}$$

Proof By using the properties of the trace and the given assumptions, the following identities are evident:

$$\begin{aligned} \mathbf{S} \cdot \mathbf{A} &= \text{tr}\left(\mathbf{S}\mathbf{A}^T\right) = \text{tr}\,(-\mathbf{S}\mathbf{A}) = -\text{tr}\,(\mathbf{S}\mathbf{A}) = -\text{tr}\,(\mathbf{A}\mathbf{S}) = -\text{tr}\left(\mathbf{A}\mathbf{S}^T\right) \\ &= -\mathbf{A} \cdot \mathbf{S} = -\mathbf{S} \cdot \mathbf{A}; \end{aligned}$$

thus

$$\mathbf{S} \cdot \mathbf{A} = -\mathbf{S} \cdot \mathbf{A} \ \Rightarrow \ \mathbf{S} \cdot \mathbf{A} = 0.$$

□

There is a simple but important corollary of the previous theorem:

Corollary 1

$$\textit{If} \quad \mathbf{S} \cdot \mathbf{A} = 0 \quad \forall\, \mathbf{A} \in \mathcal{S}kw, \ \textit{then} \ \ \mathbf{S} \in \mathcal{S}ym \tag{1.47}$$

or

$$\textit{If}\ \ \mathbf{S}\cdot\mathbf{A}=0\quad \forall\,\mathbf{S}\in \mathcal{Sym},\ \ \textit{then}\ \ \mathbf{A}\in \mathcal{Skw}. \tag{1.48}$$

Proof The proof of (1.47) can be obtained in the following way. Since the scalar product is zero for any antisymmetric matrix **A**, we can choose **A** with all elements zero except $A_{12} = -A_{21} = 1$. Thus,

$$0 = \mathbf{S}\cdot\mathbf{A} = S_{12}A_{12} + S_{21}A_{21} = S_{12} - S_{21}, \quad \Rightarrow \quad S_{12} = S_{21}.$$

Similarly choosing all elements of **A** zero except one time $A_{13} = -A_{31} = 1$ and one time $A_{23} = -A_{32} = 1$, we find that **S** is symmetric. From Theorem 1.13.1, we are sure that the scalar product is zero for any other antisymmetric matrix **A**. In a similar way, we can prove (1.48). □

1.13.2 Symmetric and Antisymmetric Parts of an Operator

There exists a simple yet important theorem:

Theorem 1.13.2 *For every operator* **L** $\in$ *Lin, there exist a symmetric operator* $\mathbf{L}^S \in$ *Sym and an antisymmetric operator* $\mathbf{L}^A \in$ *Skw such that*

$$\mathbf{L} = \mathbf{L}^S + \mathbf{L}^A. \tag{1.49}$$

The operators $\mathbf{L}^S$ and $\mathbf{L}^A$ are called the *symmetric part* and *antisymmetric part* of the operator **L**, respectively, and they are given:

$$\mathbf{L}^S = \frac{1}{2}\left(\mathbf{L}+\mathbf{L}^T\right)$$
$$\mathbf{L}^A = \frac{1}{2}\left(\mathbf{L}-\mathbf{L}^T\right).$$

The proof of this statement is simple and based on the identity

$$\mathbf{L} = \frac{1}{2}\left(\mathbf{L}+\mathbf{L}^T\right) + \frac{1}{2}\left(\mathbf{L}-\mathbf{L}^T\right) = \mathbf{L}^S + \mathbf{L}^A.$$

Hence,

$$\mathbf{L}^{ST} = \frac{1}{2}\left(\mathbf{L}+\mathbf{L}^T\right)^T = \frac{1}{2}\left(\mathbf{L}^T+\mathbf{L}\right) = \mathbf{L}^S \ \Rightarrow\ \mathbf{L}^S \in \mathcal{Sym}$$

and

$$\mathbf{L}^{AT} = \frac{1}{2}\left(\mathbf{L} - \mathbf{L}^T\right)^T = \frac{1}{2}\left(\mathbf{L}^T - \mathbf{L}\right) = -\mathbf{L}^A \quad \Rightarrow \quad \mathbf{L}^A \in \mathcal{S}kw.$$

Note that by Theorem 1.13.1, $\mathbf{L}^S$ and $\mathbf{L}^A$ are orthogonal operators.

1.14 Deviatoric and Isotropic Parts of an Operator

Let $\mathbf{S}$ be an operator with components S_{ij} $(n \times n)$ with respect to a basis $\{\mathbf{e}_i\}$. The *deviatoric part* (or *traceless part*) of $\mathbf{S}$ is defined as the linear operator $\mathbf{S}^D$ in $\mathbb{R}^n$ given by

$$\mathbf{S}^D = \mathbf{S} - \frac{1}{n}\,(\mathrm{tr}\,\mathbf{S})\,\mathbf{I}. \tag{1.50}$$

Recalling that $\mathrm{tr}\,\mathbf{I} = n$, we immediately have

$$\mathrm{tr}\,\mathbf{S}^D = 0.$$

We denote the components of $\mathbf{S}^D$ as $S_{<ij>}$, and thus (1.50) becomes in component form

$$S_{<ij>} = S_{ij} - \frac{1}{n}\,(\mathrm{tr}\,\mathbf{S})\,\delta_{ij}. \tag{1.51}$$

Equation (1.50) can also be written as

$$\mathbf{S} = \mathbf{S}^D + \mathbf{S}^I, \tag{1.52}$$

where we have defined

$$\mathbf{S}^I = \frac{1}{n}\,(\mathrm{tr}\,\mathbf{S})\,\mathbf{I}.$$

The $\mathbf{S}^D$ and $\mathbf{S}^I$ are called the *deviatoric part* and the *isotropic part* of $\mathbf{S}$, respectively.

Therefore, *an operator can always be decomposed into the sum of a deviatoric operator and an isotropic operator.*

Note that the deviatoric part and the isotropic part are orthogonal to each other, as

$$\mathbf{S}^D \cdot \mathbf{I} = \mathrm{tr}\,\mathbf{S}^D = 0.$$

Remark 3 The decomposition (1.52) holds for any operator, but it is effectively used only for symmetric operators since antisymmetric operators have an identically zero isotropic part and coincide with their deviatoric part.

1.15 Rotation Operator

We define a *rotation operator* $\mathbf{R}$ as an operator that satisfies the following conditions:

$$\mathbf{R}^T\mathbf{R} = \mathbf{I} \tag{1.53}$$

$$\det \mathbf{R} = 1. \tag{1.54}$$

From (1.53), it immediately follows that for a rotation operator $\mathbf{R}$,

$$\mathbf{R}^{-1} = \mathbf{R}^T. \tag{1.55}$$

Furthermore, from this relationship and (1.54), using the property in Eq. (1.20), we also have

$$\mathbf{R}^C = \mathbf{R}. \tag{1.56}$$

Rotation operators derive their name from the fact that they act on a vector $\mathbf{v}$ by rotating it without altering its module. In fact (see Fig. 1.5),

$$(\mathbf{Rv})^2 = \mathbf{Rv} \cdot \mathbf{Rv} = \mathbf{v} \cdot \mathbf{R}^T\mathbf{Rv} = \mathbf{v} \cdot \mathbf{Iv} = \mathbf{v} \cdot \mathbf{v} = \mathbf{v}^2.$$

Another notable property of rotation operators is that if we take two vectors $\mathbf{v}$ and $\mathbf{w}$ forming an angle θ, the transformed vectors under the rotation operator will retain the same angle θ (see Fig. 1.5):

$$\mathbf{Rv} \cdot \mathbf{Rw} = \mathbf{R}^T\mathbf{Rv} \cdot \mathbf{w} = \mathbf{v} \cdot \mathbf{w}. \tag{1.57}$$

As a consequence, applying a rotation operator $\mathbf{R}$ to the elements of an orthonormal basis $\{\mathbf{e}_i\}$ yields another orthonormal basis $\left\{\mathbf{e}'_i\right\}$:

$$\mathbf{e}'_i \cdot \mathbf{e}'_j = \mathbf{Re}_i \cdot \mathbf{Re}_j = \mathbf{e}_i \cdot \mathbf{R}^T\mathbf{Re}_j = \mathbf{e}_i \cdot \mathbf{Ie}_j = \mathbf{e}_i \cdot \mathbf{e}_j = \delta_{ij}.$$

Let us add an additional observation, valid in the case of a three-dimensional space ($n = 3$): If the basis $\{\mathbf{e}_i\}$ is *counterclockwise* (or *left-handed*),

$$\mathbf{e}_1 \times \mathbf{e}_2 \cdot \mathbf{e}_3 = 1,$$

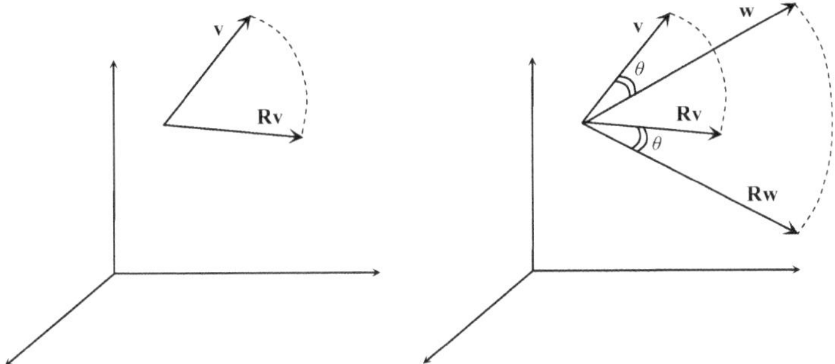

Fig. 1.5 Rotation operator

then the new basis $\{\mathbf{e}'_i\}$ obtained by transforming the elements of $\{\mathbf{e}_i\}$ through the operator $\mathbf{R}$ is also counterclockwise. In fact, taking into account Eq. (1.13),[3]

$$\mathbf{e}'_1 \times \mathbf{e}'_2 \cdot \mathbf{e}'_3 = \mathbf{R}\mathbf{e}_1 \times \mathbf{R}\mathbf{e}_2 \cdot \mathbf{R}\mathbf{e}_3 = \det \mathbf{R}\ \mathbf{e}_1 \times \mathbf{e}_2 \cdot \mathbf{e}_3 = 1.$$

From (1.53), the first of the two conditions that define a rotation operator, we have, using (1.10),

$$\det\left(\mathbf{R}^T\mathbf{R}\right) = \det \mathbf{I}$$
$$\Rightarrow \det \mathbf{R}^T \det \mathbf{R} = 1 \Rightarrow \det \mathbf{R}^2 = 1 \Rightarrow \det \mathbf{R} = \pm 1.$$

The requirement that $\det \mathbf{R} = 1$ is intended to ensure that the rotation associated with the operator $\mathbf{R}$ does not alter the symmetry of the basis. Operators that satisfy only the condition (1.53) are called *orthogonal*. Rotation operators are also called *proper orthogonal*, and in the following, to indicate that an operator is a rotation operator, we will write $\mathbf{R} \in \mathcal{R}ot$.

1.16 Orthogonal Similarity Transformations

We now want to determine the relationship between the components of a vector $\mathbf{v}$ and a matrix operator $\mathbf{A}$ expressed with respect to a counterclockwise orthonormal basis $\{\mathbf{e}_i\}$ and the analogous components expressed with respect to a basis $\{\mathbf{e}'_i\}$ obtained by transforming each component of the basis $\{\mathbf{e}_i\}$ using a rotation operator $\mathbf{R}$. Let v_i be the components of the vector $\mathbf{v}$ with respect to the basis $\{\mathbf{e}_i\}$ and v'_i be the components of the same vector with respect to the basis $\{\mathbf{e}'_i\}$. To find the

[3] The same invariance property is valid if the basis is *clockwise* (or *right-handed*).

relationship between the components v_i and v'_i, we recall that $\mathbf{e}'_i = \mathbf{R}\mathbf{e}_i$ and write

$$\mathbf{v} = v'_i \mathbf{e}'_i = v'_i \, \mathbf{R}\mathbf{e}_i .$$

Scalar multiplication of both sides by $\mathbf{e}_j$ gives

$$\mathbf{v} \cdot \mathbf{e}_j = v'_i \, \mathbf{R}\mathbf{e}_i \cdot \mathbf{e}_j ,$$

which, using (1.3) and (1.4), becomes

$$v_j = R_{ji} v'_i ,$$

which can also be written in vector form as

$$\mathbf{v} = \mathbf{R}\mathbf{v}' \tag{1.58}$$

or, thanks to the properties of the rotation operator $\mathbf{R}$,

$$\mathbf{v}' = \mathbf{R}^T \mathbf{v} \tag{1.59}$$

where we have defined

$$\mathbf{v} \equiv (v_1, v_2, \dots, v_n) , \qquad \mathbf{v}' \equiv \left(v'_1, v'_2, \dots, v'_n\right) .$$

It is important to emphasize that the symbols $\mathbf{v}$ and $\mathbf{v}'$ here represent the same vector; the fact that two different symbols have been introduced to indicate the same object is simply a matter of convenience and serves to highlight that the components of the vector are expressed with respect to two different bases.

However, there is another interpretation of the relationships (1.58) and (1.59), which fully justifies the introduction of the two symbols $\mathbf{v}$ and $\mathbf{v}'$.

We can think that, instead of rotating the basis using the rotation operator $\mathbf{R}$, the vector undergoes the reverse rotation $\mathbf{R}^{-1} = \mathbf{R}^T$, while the basis remains unchanged. The resulting vector $\mathbf{v}'$ satisfies (1.59) (see Fig. 1.6). In this case, therefore, v_i and v'_i are the components of two different vectors expressed with respect to the same basis.

Two vectors $\mathbf{v}$ *and* $\mathbf{v}'$ *that satisfy* (1.58) *and* (1.59) *are said to be related by an orthogonal similarity transformation.*[4]

Also note that the norm of a vector is an invariant quantity with respect to an orthogonal similarity transformation.

Now we determine the relationship between the components of a matrix operator $\mathbf{A}$ expressed with respect to a basis $\{\mathbf{e}_i\}$ and a basis $\left\{\mathbf{e}'_i\right\}$ obtained by transforming

[4] There exist more general similarity transformations than those considered here, but they go beyond the scope of present text.

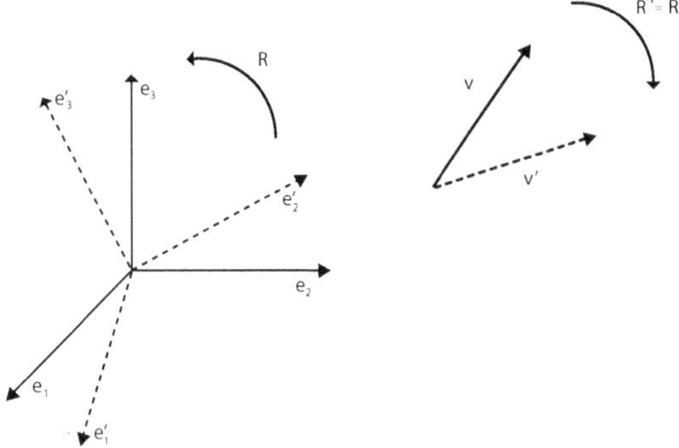

Fig. 1.6 Similarity transformations between vectors

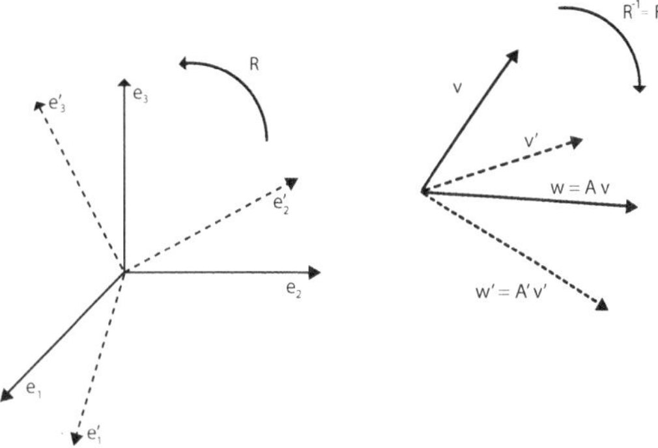

Fig. 1.7 Similarity transformations between operators

each component of the basis $\{\mathbf{e}_i\}$ using the rotation operator $\mathbf{R}$. Let $\mathbf{v}$ and $\mathbf{w}$ be two vectors such that

$$\mathbf{w} = \mathbf{A}\mathbf{v},$$

and let $\mathbf{v}'$ and $\mathbf{w}'$ be the respective vectors obtained by the similarity transformation

$$\mathbf{v}' = \mathbf{R}^T\mathbf{v}$$
$$\mathbf{w}' = \mathbf{R}^T\mathbf{w}.$$

The operator $\mathbf{A}'$ obtained by similarity transformation of $\mathbf{A}$ will therefore satisfy (see Fig. 1.7)

$$\mathbf{w}' = \mathbf{A}'\mathbf{v}'. \tag{1.60}$$

Thus, we have

$$\mathbf{w}' = \mathbf{R}^T\mathbf{w} = \mathbf{R}^T\mathbf{A}\mathbf{v} = \mathbf{R}^T\mathbf{A}\mathbf{R}\mathbf{v}', \tag{1.61}$$

and from (1.60) and (1.61), it immediately follows that

$$\mathbf{A}' = \mathbf{R}^T \mathbf{A}\mathbf{R}, \qquad A'_{ij} = R_{ki} A_{kt} R_{tj} \tag{1.62}$$

$$\mathbf{A} = \mathbf{R}\mathbf{A}'\mathbf{R}^T, \qquad A_{ij} = R_{ik} A'_{kt} R_{jt}. \tag{1.63}$$

Operators that satisfy the above relations are said to be related by an orthogonal similarity transformation (or simply a similarity transformation).

Note that isotropic operators (1.6) remain unchanged under a similarity transformation. In fact, if $\mathbf{A} = \lambda \mathbf{I}$, we have

$$\mathbf{A}' = \mathbf{R}^T (\lambda\, \mathbf{I})\, \mathbf{R} = \lambda\, \mathbf{R}^T \mathbf{R} = \lambda\, \mathbf{I} = \mathbf{A}.$$

Isotropic operators have components that do not depend on the choice of the basis.

1.16.1 Principal Invariants of an Operator

In the case of $n \times n$ matrix operators, it is possible to identify n independent scalars that are invariant under a similarity transformation.

In the most interesting case for us, $n = 3$, these quantities are specifically the trace, the trace of the cofactor operator, and the determinant:

$$I_1 = \operatorname{tr} \mathbf{A}, \qquad I_2 = \operatorname{tr} \mathbf{A}^C, \qquad I_3 = \det \mathbf{A}. \tag{1.64}$$

Naturally, any function of these three invariants (known as *principal invariants*) is also an invariant quantity.

Let us verify the invariance of I_1, I_2, and I_3 by recalling the properties of the trace, in particular Eq. (1.8), and the determinant:

- *Invariance of the trace* (I_1, *first principal invariant*):

$$I_1 = \operatorname{tr} \mathbf{A}' = \operatorname{tr} \left(\mathbf{R}^T \mathbf{A}\mathbf{R}\right) = \operatorname{tr} \left(\mathbf{A}\mathbf{R}\mathbf{R}^T\right) = \operatorname{tr} \left(\mathbf{A}\mathbf{I}\right) = \operatorname{tr} \mathbf{A}.$$

- *Invariance of the trace of the cofactor operator* (I_2, *second principal invariant*):

$$\begin{aligned} I_2 = \operatorname{tr} \mathbf{A}'^C &= \operatorname{tr} \left(\mathbf{R}^T \mathbf{A}\mathbf{R}\right)^C = \operatorname{tr} \left(\mathbf{R}^{TC} \mathbf{A}^C \mathbf{R}^C\right) \\ &= \operatorname{tr} \left(\mathbf{R}^T \mathbf{A}^C \mathbf{R}\right) = \operatorname{tr} \left(\mathbf{A}^C \mathbf{R}\mathbf{R}^T\right) \\ &= \operatorname{tr} \left(\mathbf{A}^C \mathbf{I}\right) = \operatorname{tr} \mathbf{A}^C. \end{aligned}$$

- *Invariance of the determinant* (I_3, *third principal invariant*):

$$I_3 = \det \mathbf{A}' = \det \left(\mathbf{R}^T \mathbf{A} \mathbf{R}\right) = \det \mathbf{R}^T \det \mathbf{A} \det \mathbf{R} = \det \mathbf{A}.$$

1.17 Eigenvalues and Eigenvectors of an Operator

Given an operator $\mathbf{A} \in \mathcal{L}in$ of order n, it is of interest to find nonzero vectors $\mathbf{v}$ (if they exist) such that when the operator $\mathbf{A}$ is applied to them, the resulting vector is parallel to the vector $\mathbf{v}$ itself. In other words, we are looking for vectors $\mathbf{v}$ for which the following equation holds:

$$\mathbf{A}\mathbf{v} = \lambda \mathbf{v} \tag{1.65}$$

for some λ. Equation (1.65) can be rewritten as

$$(\mathbf{A} - \lambda \mathbf{I})\mathbf{v} = 0, \tag{1.66}$$

which represents a homogeneous system of linear algebraic equations. It is well known that such a system has a nontrivial solution $\mathbf{v} \neq 0$ if and only if

$$\det(\mathbf{A} - \lambda \mathbf{I}) = 0. \tag{1.67}$$

Equation (1.67) represents a polynomial of degree n in the variable λ (*characteristic polynomial*).

The roots $\lambda^{(i)}$ of the characteristic polynomial are called *eigenvalues* of the operator $\mathbf{A}$, and the corresponding vectors $\mathbf{v}^{(i)}$ are called *eigenvectors*. For the purpose of this book, we will focus on the case where the eigenvalues and eigenvectors are real. It is important to note that, as evident from (1.66), eigenvectors are defined up to a multiplicative scalar, which means that if $\mathbf{v}$ is an eigenvector of the operator $\mathbf{A}$, then any vector that has the same direction as $\mathbf{v}$ will also be an eigenvector. The directions identified by the real eigenvectors are called the *principal axes* of the operator $\mathbf{A}$.

In the case of $n = 3$, it can be verified that the characteristic polynomial takes the form:

$$\lambda^3 - I_1 \lambda^2 + I_2 \lambda - I_3 = 0. \tag{1.68}$$

Note that the coefficients of the characteristic polynomial are equal, up to alternate sign, to the principal invariants (hence their name), and therefore, the eigenvalues are also invariant under any similarity transformation. This property holds true for any n.

If $\mathbf{A}$ is non-singular, the inverse operator $\mathbf{A}^{-1}$ has the same eigenvectors as $\mathbf{A}$ and eigenvalues $1/\lambda$. This is evident by applying $\mathbf{A}^{-1}$ to Eq. (1.65):

$$\mathbf{A}^{-1}\mathbf{v} = \frac{1}{\lambda}\mathbf{v}. \tag{1.69}$$

1.17.1 Eigenvalues and Invariants of Operator Powers

Let $\mathbf{A}^m$ denote the m-th power operator of $\mathbf{A}$ (with m being a positive integer):

$$\mathbf{A}^m = \underbrace{\mathbf{A}\mathbf{A}\cdots\mathbf{A}}_{m \text{ times}}.$$

From (1.66), we have

$$\mathbf{A}^m\mathbf{v} = \lambda^m\mathbf{v}. \tag{1.70}$$

Hence, if $\mathbf{A}$ has eigenvalue λ and eigenvector $\mathbf{v}$, then the power operator $\mathbf{A}^m$ has the same eigenvectors $\mathbf{v}$ as $\mathbf{A}$ and eigenvalues λ^m.

The following properties can be easily verified (see (1.23) and (1.24)):

$$(\mathbf{A}^m)^T = (\mathbf{A}^T)^m; \quad (\mathbf{A}^m)^{-1} = (\mathbf{A}^{-1})^m. \tag{1.71}$$

From the trace property, it follows that for a given operator $\mathbf{A} \in \mathbb{R}^n$, the traces of all the powers of $\mathbf{A}$ are independent invariants:

$$J_m = \operatorname{tr} \mathbf{A}^m, \qquad 1 \le m \le n. \tag{1.72}$$

Consider J_m as a function of the n^2 variables A_{ij} taking values in $\mathbb{R}$. Taking partial derivatives with respect to each variable, we obtain an $n \times n$ matrix. We have

$$\frac{\partial J_1}{\partial A_{lm}} = \frac{\partial A_{ii}}{\partial A_{lm}} = \delta_{il}\delta_{im} = \delta_{lm} \Longrightarrow \frac{\partial J_1}{\partial \mathbf{A}} = \mathbf{I}$$

$$\frac{\partial J_2}{\partial A_{lm}} = \frac{\partial A_{ij}A_{ji}}{\partial A_{lm}} = \delta_{il}\delta_{jm}A_{ji} + A_{ij}\delta_{jl}\delta_{im} = 2A_{ml} \Longrightarrow \frac{\partial J_2}{\partial \mathbf{A}} = 2\mathbf{A}^T,$$

and it can be easily seen that, in general, the identity holds:

$$\frac{\partial J_m}{\partial \mathbf{A}} = m\left(\mathbf{A}^T\right)^{m-1} \quad \forall\ 1 \le m \le n. \tag{1.73}$$

1.17.2 Eigenvalues and Eigenvectors for Symmetric Operators

In the case of a symmetric operator **A**, the following important considerations are applicable:

- If an operator **A** is symmetric, then distinct eigenvalues correspond to mutually orthogonal eigenvectors. Let **u** and **v** be two eigenvectors of **A** with eigenvalues λ and μ, respectively; we have

$$\mathbf{A}\mathbf{u} \cdot \mathbf{v} = \lambda\, \mathbf{u} \cdot \mathbf{v},$$

$$\mathbf{A}\mathbf{u} \cdot \mathbf{v} = \mathbf{A}^T \mathbf{v} \cdot \mathbf{u} = \mathbf{A}\mathbf{v} \cdot \mathbf{u} = \mu\, \mathbf{u} \cdot \mathbf{v}.$$

  By subtracting the two expressions, we obtain $(\lambda - \mu)\, \mathbf{u} \cdot \mathbf{v} = 0$. Since $\lambda \neq \mu$, this implies that **u** and **v** must be orthogonal.
- A symmetric operator **A** has all real eigenvalues. If λ is an eigenvalue of **A** with the corresponding eigenvector **v**, then

$$\mathbf{A}\mathbf{v} = \lambda \mathbf{v}.$$

  If λ is a complex number, the conjugate of λ, denoted by $\bar{\lambda}$, is also an eigenvalue of the operator **A**, and we would have

$$\mathbf{A}\bar{\mathbf{v}} = \bar{\lambda}\bar{\mathbf{v}},$$

  where $\bar{\mathbf{v}}$ is the complex conjugate of **v**. Assuming $\lambda \neq \bar{\lambda}$, as just shown, the vectors **v** and $\bar{\mathbf{v}}$ must be mutually orthogonal. Therefore,

$$\mathbf{v} \cdot \mathbf{v} = 0 \quad \Rightarrow \quad |\mathbf{v}|^2 = 0$$

  contrary to the hypothesis that every eigenvector of **A** is a nonzero solution of equation (1). Hence, it can be deduced that λ cannot be different from $\bar{\lambda}$, that is, λ must be real.
- If matrix **A** has a double eigenvalue with two linearly independent eigenvectors **u** and **v**, then every vector in the plane spanned by **u** and **v** is also an eigenvector. Let $\lambda_1 = \lambda_2 = \lambda$ be the two coincident eigenvalues, and let **u** and **v** be the associated eigenvectors. We have

$$\begin{aligned} \mathbf{A}(\alpha\mathbf{u} + \beta\mathbf{v}) &= \alpha\mathbf{A}\mathbf{u} + \beta\mathbf{A}\mathbf{v} \\ &= \alpha\lambda\mathbf{u} + \beta\lambda\mathbf{v} \\ &= \lambda(\alpha\mathbf{u} + \beta\mathbf{v}). \end{aligned}$$

This shows that any linear combination of the eigenvectors $\mathbf{u}$ and $\mathbf{v}$ is still an eigenvector of $\mathbf{A}$ corresponding to the double eigenvalue. This property extends to the case of an eigenvalue with multiplicity m where $2 \leq m \leq n$.

- For a non-singular 3×3 symmetric matrix $\mathbf{A}$, if $\mathbf{u}$ and $\mathbf{v}$ are eigenvectors, then the cross product vector $\mathbf{u} \times \mathbf{v}$ is also an eigenvector. Let λ and μ be the eigenvalues corresponding to the eigenvectors $\mathbf{u}$ and $\mathbf{v}$, respectively. We have

$$\mathbf{A}\mathbf{u} = \lambda\mathbf{u}, \quad \mathbf{A}\mathbf{v} = \mu\mathbf{v}. \tag{1.74}$$

Using the identity (1.28) for the case $n = 3$

$$(\mathbf{A}^C)^C = (\det\mathbf{A})\mathbf{A}, \tag{1.75}$$

we can write, based on (1.75) and (1.42),

$$\mathbf{A}(\mathbf{u} \times \mathbf{v}) = \frac{1}{\det\mathbf{A}}(\mathbf{A}^C)^C(\mathbf{u} \times \mathbf{v}) = \frac{1}{\det\mathbf{A}}\mathbf{A}^C\mathbf{u} \times \mathbf{A}^C\mathbf{v}.$$

Taking into account (1.22), the symmetry of $\mathbf{A}$, the fact that $\mathbf{u}$ and $\mathbf{v}$ are eigenvectors of $\mathbf{A}$ (1.74) and (1.69), we have

$$\begin{aligned}\mathbf{A}(\mathbf{u} \times \mathbf{v}) &= \frac{1}{\det\mathbf{A}}\,\mathbf{A}^C\mathbf{u} \times \mathbf{A}^C\mathbf{v} \\ &= \det\mathbf{A}\ \mathbf{A}^{-1T}\mathbf{u} \times \mathbf{A}^{-1T}\mathbf{v} \\ &= \det\mathbf{A}\ \mathbf{A}^{-1}\mathbf{u} \times \mathbf{A}^{-1}\mathbf{v} \\ &= \frac{\det\mathbf{A}}{\lambda\mu}\,\mathbf{u} \times \mathbf{v}.\end{aligned}$$

This shows that $\mathbf{u} \times \mathbf{v}$ is an eigenvector of $\mathbf{A}$ with the eigenvalue $\det\mathbf{A}/(\lambda\mu)$.

1.17.3 Diagonalization of an Operator

A linear operator $\mathbf{A}$ is said to be *diagonalizable* (under orthogonal similarities) if there exists an orthonormal basis $\{\mathbf{e}_i\}$ in which its associated matrix takes the diagonal form. It can be easily verified that if $\mathbf{A}$ is diagonalizable, then the basis with respect to which $\mathbf{A}$ takes a diagonal form is composed of eigenvectors of $\mathbf{A}$, and the elements appearing on the diagonal are the eigenvalues of $\mathbf{A}$.

Theorem 1.17.1 *All and only the symmetric operators are diagonalizable under orthogonal similarities.*

Proof The fact that a symmetric operator is diagonalizable directly follows from what has been seen above. In fact, if the eigenvalues are distinct, the eigenvectors are mutually orthogonal. In the case where two or more eigenvalues coincide, there are infinitely many orthogonal bases consisting of eigenvectors. Note that a diagonalizable operator, expressed in the basis with respect to which it takes a diagonal form, is obviously symmetric. On the other hand, if an operator is symmetric in one basis, it remains symmetric even after any orthogonal similarity transformation. Indeed, given the similarity transformation represented by the rotation operator $\mathbf{R}$:

$$\mathbf{A}' = \mathbf{R}^T\mathbf{A}\mathbf{R},$$

we have

$$\mathbf{A}'^T = \left(\mathbf{R}^T\mathbf{A}\mathbf{R}\right)^T = \mathbf{R}^T\mathbf{A}^T\mathbf{R} = \mathbf{R}^T\mathbf{A}\mathbf{R} = \mathbf{A}'.$$

□

1.17.4 Hamilton–Cayley Theorem

We consider first the case of dimension $n = 3$ and symmetric operators. We have the important Hamilton–Cayley theorem:

Theorem 1.1 *Given an operator* $\mathbf{A}$, *there exists the following identity between the power of an operator:*

$$\mathbf{A}^3 - I_1\mathbf{A}^2 + I_2\mathbf{A} - I_3\mathbf{I} = 0. \tag{1.76}$$

Proof Let us multiply (1.68) by the eigenvector $\mathbf{v}$ corresponding to eigenvalue λ and consider (1.70):

$$\left(\mathbf{A}^3 - I_1\mathbf{A}^2 + I_2\mathbf{A} - I_3\mathbf{I}\right)\mathbf{v} = 0.$$

Since this equality must hold for every eigenvector and the eigenvectors of a symmetric operator are linearly independent and orthogonal to each other, we necessarily obtain (1.76). □

On the other hand, symmetric operators are diagonalizable, so if we take the eigenvectors as the basis, (1.76) is equivalent to (1.68).

The Hamilton–Cayley theorem (1.76), which holds for any operator, is of fundamental importance as it allows us to derive all powers of an operator $\mathbf{A}^m$ with $m \geqslant 3$ *as a combination of* $\mathbf{A}^2$, $\mathbf{A}$, and $\mathbf{I}$.

In fact, from (1.76), we can obtain $\mathbf{A}^3$, and by subsequently multiplying by $\mathbf{A}$, we can derive $\mathbf{A}^4$, and so on.

Moreover, assuming $\mathbf{A}$ is non-singular, by multiplying (1.76) by $\mathbf{A}^{-1}$, we obtain an important relationship between the inverse and the cofactor matrices of $\mathbf{A}$ with $\mathbf{A}^2$, $\mathbf{A}$, and $\mathbf{I}$ (taking into account $(1.71)_1$):

$$\mathbf{A}^{-1} = \frac{1}{I_3}\left(\mathbf{A}^2 - I_1\mathbf{A} + I_2\mathbf{I}\right) \tag{1.77}$$

$$\mathbf{A}^C = (\mathbf{A}^T)^2 - I_1\mathbf{A}^T + I_2\mathbf{I}. \tag{1.78}$$

The Hamilton–Cayley theorem, as well as the characteristic polynomial, extends to the case of any dimension n:

$$\mathbf{A}^n + \sum_{j=1}^{n}(-1)^j I_j \mathbf{A}^{n-j} = 0; \qquad \lambda^n + \sum_{j=1}^{n}(-1)^j I_j \lambda^{n-j} = 0,$$

where I_j $(j = 1, 2, \cdots, n)$ denotes the n principal invariants and $\mathbf{A}^0 = \mathbf{I}$. In the general case, given an operator $\mathbf{A}$, the theorem provides all powers $\mathbf{A}^m$ with $m \geqslant n$, $\mathbf{A}^{-1}$ (if it exists), and $\mathbf{A}^C$ as a linear combination of the powers $\mathbf{A}^s$ $(s = 0, 1, \cdots, n-1)$.

1.17.5 Relations Between Invariants and Derivatives of Principal Invariants in the Case $n = 3$

In the most interesting case of dimension $n = 3$, it is useful to know the derivatives of the principal invariants of an operator with respect to the operator itself. To this end, we recall that we know the derivatives of the invariants J_m (see (1.73)), so we just need to find the relations between I_1, I_2, and I_3 and J_1, J_2, and J_3 (see (1.64) and (1.72)). Obviously,

$$J_1 = I_1,$$

while to find the relation with J_2, we take the trace of both sides of (1.78) and immediately obtain

$$J_2 = I_1^2 - 2I_2.$$

Finally, by taking the trace of (1.76), we obtain

$$J_3 = I_1^3 + 3\,(I_3 - I_1 I_2)\,.$$

It is possible to invert the previous relations, obtaining

$$\begin{cases} I_1 = J_1 \\ I_2 = \frac{1}{2}\left(J_1^2 - J_2\right) \\ I_3 = \frac{1}{3}\left(J_3 - J_1^3\right) + \frac{J_1}{2}\left(J_1^2 - J_2\right). \end{cases} \tag{1.79}$$

From (1.79), (1.73), and (1.78), we immediately obtain the desired formulas that will be useful in the upcoming chapters:

$$\frac{\partial I_1}{\partial \mathbf{A}} = \mathbf{I}, \qquad \frac{\partial I_2}{\partial \mathbf{A}} = (\operatorname{tr} \mathbf{A})\, \mathbf{I} - \mathbf{A}^T, \qquad \frac{\partial I_3}{\partial \mathbf{A}} = \mathbf{A}^C. \tag{1.80}$$

1.18 Tensor Product

We define the *tensor product* $\mathbf{T}$ between two vectors $\mathbf{u}$ and $\mathbf{v}$ as

$$\mathbf{T} = \mathbf{u} \otimes \mathbf{v},$$

where the components of $\mathbf{T}$ are given by

$$T_{ij} = u_i v_j.$$

The tensor product satisfies the following identities:

- $\operatorname{tr}(\mathbf{u} \otimes \mathbf{v}) = \mathbf{u} \cdot \mathbf{v}$.
- $\mathbf{u} \otimes \mathbf{v} = (\mathbf{v} \otimes \mathbf{u})^T$.
- $\det(\mathbf{u} \otimes \mathbf{v}) = 0$.
- $(\mathbf{u} \otimes \mathbf{v})\, \mathbf{w} = (\mathbf{v} \cdot \mathbf{w})\, \mathbf{u}$.

The first two identities are immediately verifiable. Regarding the third identity, using the properties of determinants, we can factor out a common factor in a row or a column and observe that the determinant is zero if two rows or columns are proportional to each other. Hence, we have

$$\det \mathbf{T} = \det \begin{pmatrix} u_1 v_1 & u_1 v_2 & \dots & u_1 v_n \\ u_2 v_1 & u_2 v_2 & \dots & u_2 v_n \\ \dots & \dots & \dots & \dots \\ u_n v_1 & u_n v_2 & \dots & u_n v_n \end{pmatrix}$$

$$= u_1 \det \begin{pmatrix} v_1 & v_2 & \dots & v_n \\ u_2 v_1 & u_2 v_2 & \dots & u_2 v_n \\ \dots & \dots & \dots & \dots \\ u_n v_1 & u_n v_2 & \dots & u_n v_n \end{pmatrix}$$

$$= u_1 u_2 \det \begin{pmatrix} v_1 & v_2 & \dots & v_n \\ v_1 & v_2 & \dots & v_n \\ \dots & \dots & \dots & \dots \\ u_n v_1 & u_n v_2 & \dots & u_n v_n \end{pmatrix}$$

$$= 0.$$

The last property can be demonstrated as follows. Let $\mathbf{z} = (\mathbf{u} \otimes \mathbf{v})\,\mathbf{w}$. We have

$$z_i = ((\mathbf{u} \otimes \mathbf{v})\,\mathbf{w})_i = (\mathbf{u} \otimes \mathbf{v})_{ij}\, w_j = u_i v_j w_j = (\mathbf{v} \cdot \mathbf{w})\, u_i .$$

Therefore,

$$z_i = (\mathbf{v} \cdot \mathbf{w})\, u_i \quad \Rightarrow \quad \mathbf{z} = (\mathbf{v} \cdot \mathbf{w})\,\mathbf{u}.$$

1.18.1 Semi-Cartesian Representation of an Operator

The scalar product allows us to represent an operator in terms of its basis, similar to the semi-Cartesian representation of a vector. While for vectors we have the well-known expression

$$\mathbf{v} = v_i \mathbf{e}_i ,$$

it is straightforward to verify that, thanks to the previously defined tensor product, we can write for any operator

$$\mathbf{A} = A_{ij} \mathbf{e}_i \otimes \mathbf{e}_j . \tag{1.81}$$

To demonstrate this, we can consider the generic components lm on both sides and recall that

$$\left(\mathbf{e}_i \otimes \mathbf{e}_j\right)_{lm} = \delta_{il} \delta_{mj} .$$

This identity verifies Eq. (1.81).

1.18.2 *Eigenvalues and Eigenvectors of a Tensor Product for* n = 3

Considering that for a tensor product $\mathbf{T} = \mathbf{u} \otimes \mathbf{v}$, we have $I_1 = \mathbf{u} \cdot \mathbf{v}$, $I_2 = I_3 = 0$, we obtain from (1.68)

$$\lambda_1 = \lambda_2 = 0, \quad \lambda_3 = \mathbf{u} \cdot \mathbf{v},$$

while from (1.65)

$$\mathbf{Tw} = \lambda \mathbf{w} \quad \Longleftrightarrow (\mathbf{v} \cdot \mathbf{w})\, \mathbf{u} = \lambda \mathbf{w}.$$

Thus, the eigenvectors of $\mathbf{T}$ are $\mathbf{w}_1$ and $\mathbf{w}_2$ orthogonal to $\mathbf{v}$, while $\mathbf{w}_3$ will be parallel to $\mathbf{u}$:

$$\mathbf{w}_1 = \boldsymbol{\omega}_1 \times \mathbf{v}, \qquad \mathbf{w}_2 = \boldsymbol{\omega}_2 \times \mathbf{v}, \qquad \mathbf{w}_3 = \mathbf{u}, \qquad (\boldsymbol{\omega}_1 \cdot \mathbf{v} = \boldsymbol{\omega}_2 \cdot \mathbf{v} = 0).$$

Note that in the case where $\mathbf{u} \cdot \mathbf{v} = 0$, the zero eigenvalue has multiplicity 3, but in this case, the eigenvectors are not linearly independent.

1.19 Definite Sign Operators

An operator $\mathbf{A}$ is said to be *positive definite* or *negative definite* if the following conditions hold, respectively:

$$\mathbf{Av} \cdot \mathbf{v} > 0 \;\; \forall \mathbf{v} \neq 0, \qquad \mathbf{Av} \cdot \mathbf{v} = 0 \;\; \Leftrightarrow \;\; \mathbf{v} = 0,$$

$$\mathbf{Av} \cdot \mathbf{v} < 0 \;\; \forall \mathbf{v} \neq 0, \qquad \mathbf{Av} \cdot \mathbf{v} = 0 \;\; \Leftrightarrow \;\; \mathbf{v} = 0.$$

An operator $\mathbf{A}$ is said to be *positive semidefinite* or *negative semidefinite* if the following conditions hold, in order:

$$\mathbf{Av} \cdot \mathbf{v} \geq 0 \;\; \forall \mathbf{v}, \quad \exists\, \mathbf{v} \neq 0 : \;\; \mathbf{Av} \cdot \mathbf{v} = 0,$$

$$\mathbf{Av} \cdot \mathbf{v} \leq 0 \;\; \forall \mathbf{v}, \quad \exists\, \mathbf{v} \neq 0 : \;\; \mathbf{Av} \cdot \mathbf{v} = 0.$$

From the previous definitions, it is immediately seen that positive definite and negative definite operators transform a generic vector $\mathbf{v}$ into a vector $\mathbf{Av}$ that forms an angle, respectively, $\alpha < \pi/2$ and $\alpha > \pi/2$, with $\mathbf{v}$.

Similarly, it can be observed that positive semidefinite and negative semidefinite operators transform a generic vector $\mathbf{v}$ into a vector that forms an angle $\alpha \leq \pi/2$ or $\alpha \geq \pi/2$ with $\mathbf{v}$, respectively (see Fig. 1.8).

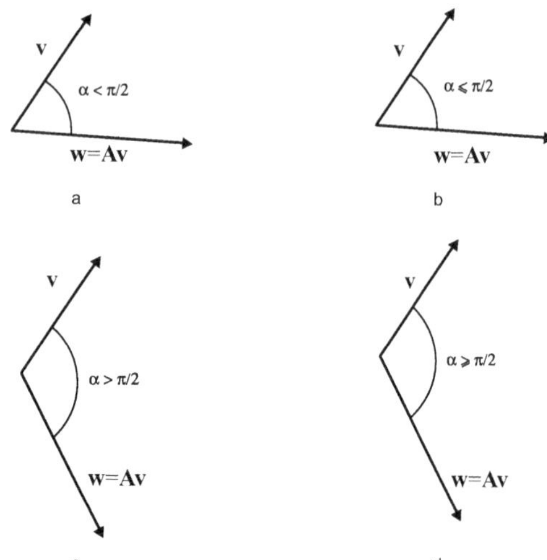

Fig. 1.8 Definite sign operators. (**a**) Positive Definite Operator. (**b**) Positive Semi-Definite Operator. (**c**) Negative Definite Operator. (**d**) Negative Semi-Definite Operator

Remark 4 The analysis of the sign of a nonsymmetric operator **F** can be reduced to the sign analysis of its symmetric part $\mathbf{F}^S$. This is due to the decomposition (1.49) and the fact that every antisymmetric operator transforms any vector **v** into a vector orthogonal to **v** (see (1.44)):

$$\mathbf{F}\mathbf{v}\cdot\mathbf{v} = \left(\mathbf{F}^S + \mathbf{F}^A\right)\mathbf{v}\cdot\mathbf{v} = \mathbf{F}^S\mathbf{v}\cdot\mathbf{v}.$$

Therefore, concerning the sign definiteness of an operator from now we consider without loss of generality only symmetric operators.

The following properties hold for definite sign operators:

- *The eigenvalues of a symmetric positive definite operator* **A** *are all strictly positive.* As $\mathbf{A}\mathbf{v}\cdot\mathbf{v}$ is a scalar and therefore independent of the choice of basis, we can choose the eigenvectors of **A**, as basis for which **A** is diagonal, and we have

$$\mathbf{A}\mathbf{v}\cdot\mathbf{v} = \lambda_1 v_1^2 + \lambda_2 v_2^2 + \lambda_3 v_3^2 > 0 \quad \forall\mathbf{v} \neq 0. \tag{1.82}$$

  It is evident that all $\lambda_i > 0$ $(i = 1, 2, 3)$ must be positive, considering that the inequality must hold for all nonzero vectors **v** including the vector of the basis $\mathbf{v} = \mathbf{e}_1 \equiv (1, 0, 0)$, $\mathbf{v} = \mathbf{e}_2 \equiv (0, 1, 0)$, $\mathbf{v} = \mathbf{e}_3 \equiv (0, 0, 1)$ for which we have from (1.82), respectively, $\lambda_1 > 0$, $\lambda_2 > 0$, $\lambda_3 > 0$. Then, for any other $\mathbf{v} \neq 0$, $\mathbf{A}\mathbf{v}\cdot\mathbf{v} > 0$ and will be zero if and only if $\mathbf{v} = 0$. The proof is valid also in dimension n.
- *The eigenvalues of a symmetric negative definite operator* **A** *are all strictly negative.* The proof is similar to the one of the previous case.
- *An operator* **A** *defined by the relation* $\mathbf{A} = \mathbf{F}^T\mathbf{F}$, *with* **F** *being a non-singular operator, is symmetric and positive definite.* The symmetry of the operator **A** can

be verified as follows:

$$\mathbf{A}^T = \left(\mathbf{F}^T\mathbf{F}\right)^T = \mathbf{F}^T\mathbf{F} = \mathbf{A}. \tag{1.83}$$

It can also be easily verified that $\mathbf{A}$ is positive definite. Indeed,

$$\mathbf{A}\mathbf{v}\cdot\mathbf{v} = \left(\mathbf{F}^T\mathbf{F}\right)\mathbf{v}\cdot\mathbf{v} = \mathbf{F}^T\left(\mathbf{F}\mathbf{v}\right)\cdot\mathbf{v} = \mathbf{F}\mathbf{v}\cdot\mathbf{F}\mathbf{v} = \left(\mathbf{F}\mathbf{v}\right)^2. \tag{1.84}$$

From this expression, it is evident that since $\det\mathbf{F} \neq 0$ by assumption, $\mathbf{A}$ is positive definite. In fact $\mathbf{A}\mathbf{v}\cdot\mathbf{v} = 0$ implies $(\mathbf{F}\mathbf{v})^2$, i.e., $\mathbf{F}\mathbf{v} = 0$, but as $\det\mathbf{F} \neq 0$ follows $\mathbf{v} = 0$. For all other non-null vectors from (1.84), we have $\mathbf{A}\mathbf{v}\cdot\mathbf{v} > 0$.

1.19.1 Sylvester's Criterion

To verify whether a symmetric operator $\mathbf{A}$ is positive definite, positive semidefinite, negative definite, or negative semidefinite, the following criteria can be used.

Let Δ_k be the determinant given by

$$\Delta_k = \det\begin{pmatrix} A_{11} & A_{12} & \dots & A_{1k} \\ A_{21} & A_{22} & \dots & A_{2k} \\ \dots & \dots & \dots & \dots \\ A_{k1} & A_{k2} & \dots & A_{kk} \end{pmatrix}.$$

The following results hold:

- The operator $\mathbf{A}$ is positive definite if and only if $\Delta_k > 0$ for $k = 1, 2, \dots, n$.
- The operator $\mathbf{A}$ is negative definite if and only if $(-1)^k\Delta_k > 0$ for $k = 1, 2, \dots, n$.

Let $\hat{\Delta}_k$ be the determinants given by

$$\hat{\Delta}_k = \det\begin{pmatrix} A_{i_1i_1} & A_{i_1i_2} & \dots & A_{i_1i_k} \\ A_{i_2i_1} & A_{i_2i_2} & \dots & A_{i_2i_k} \\ \dots & \dots & \dots & \dots \\ A_{i_ki_1} & A_{i_ki_2} & \dots & A_{i_ki_k} \end{pmatrix},$$

where the indices $1 \leq i_1 < i_2 < \ldots < i_k \leq n$ vary. Then the following results hold:

- The operator $\mathbf{A}$ is positive semidefinite if and only if all $\hat{\Delta}_k$ for $k = 1, 2, \ldots, n$ are nonnegative, and at least one of them is zero.
- The operator $\mathbf{A}$ is negative semidefinite if and only if all even-order $\hat{\Delta}_k$ are nonnegative and all odd-order $\hat{\Delta}_k$ are nonpositive, and at least one of them is zero.

For example, for a symmetric operator $\mathbf{A}$ in a three-dimensional space, the above criteria reduce to the following:

- The operator $\mathbf{A}$ is positive definite if and only if

$$\Delta_1 = A_{11} > 0, \quad \Delta_2 = \det\begin{pmatrix} A_{11} & A_{12} \\ A_{12} & A_{22} \end{pmatrix} > 0, \quad \Delta_3 = \det \mathbf{A} > 0.$$

- The operator $\mathbf{A}$ is negative definite if and only if

$$\Delta_1 = A_{11} < 0, \quad \Delta_2 = \det\begin{pmatrix} A_{11} & A_{12} \\ A_{12} & A_{22} \end{pmatrix} > 0, \quad \Delta_3 = \det \mathbf{A} < 0.$$

- The operator $\mathbf{A}$ is positive semidefinite if and only if

$$\hat{\Delta}_1 \geq 0: \quad A_{11} \geq 0, \quad A_{22} \geq 0, \quad A_{33} \geq 0$$

$$\hat{\Delta}_2 \geq 0: \quad \det\begin{pmatrix} A_{11} & A_{12} \\ A_{12} & A_{22} \end{pmatrix} \geq 0, \quad \det\begin{pmatrix} A_{11} & A_{13} \\ A_{13} & A_{33} \end{pmatrix} \geq 0, \quad \det\begin{pmatrix} A_{22} & A_{23} \\ A_{23} & A_{33} \end{pmatrix} \geq 0$$

$$\hat{\Delta}_3 \geq 0: \quad \det \mathbf{A} \geq 0,$$

  and at least one of them is equal to zero.
- The operator $\mathbf{A}$ is negative semidefinite if and only if

$$\hat{\Delta}_1 \leq 0: \quad A_{11} \leq 0, \quad A_{22} \leq 0, \quad A_{33} \leq 0$$

$$\hat{\Delta}_2 \geq 0: \quad \det\begin{pmatrix} A_{11} & A_{12} \\ A_{12} & A_{22} \end{pmatrix} \geq 0, \quad \det\begin{pmatrix} A_{11} & A_{13} \\ A_{13} & A_{33} \end{pmatrix} \geq 0, \quad \det\begin{pmatrix} A_{22} & A_{23} \\ A_{23} & A_{33} \end{pmatrix} \geq 0$$

$$\hat{\Delta}_3 \leq 0: \quad \det \mathbf{A} \leq 0,$$

  and at least one of them is equal to zero.

For symmetric operators, we will use the symbol $\mathbf{S} \in \mathcal{Sym}^+$ to indicate that $\mathbf{S}$ is positive definite and $\mathbf{S} \in \mathcal{Sym}^-$ to indicate negative definite symmetric operators.

1.19.2 Square Root Operator of a Definite Positive Operator

Let $\mathbf{S} \in \mathcal{S}ym^+$ be a symmetric operator, and let S_{ij} be its components with respect to a given orthonormal basis $\{\mathbf{e}_i\}$. Due to the properties mentioned above, there exists a set of orthonormal eigenvectors $\{\mathbf{d}_i\}$ such that $\mathbf{S}$ can be diagonalized with all positive eigenvalues. Denoting $\mathbf{R}$ as the rotation operator that transforms $\mathbf{e}_i$ to $\mathbf{d}_i = \mathbf{R}\mathbf{e}_i$, we have

$$\mathbf{S} = \mathbf{R}\mathbf{S}'\mathbf{R}^T, \quad \mathbf{S}' \equiv \begin{pmatrix} \lambda_1 & 0 & \dots & 0 \\ 0 & \lambda_2 & \dots & 0 \\ \dots & \dots & \dots & \dots \\ 0 & 0 & \dots & \lambda_n \end{pmatrix}.$$

With respect to the basis $\{\mathbf{d}_i\}$, we can define the matrix

$$\mathbf{C}' \equiv \begin{pmatrix} \sqrt{\lambda_1} & 0 & \dots & 0 \\ 0 & \sqrt{\lambda_2} & \dots & 0 \\ \dots & \dots & \dots & \dots \\ 0 & 0 & \dots & \sqrt{\lambda_n} \end{pmatrix}, \tag{1.85}$$

as all λ_i are positive. It is evident that $\mathbf{C}'^2 = \mathbf{S}'$.

We define the square root operator of $\mathbf{S}$ as the operator $\mathbf{C} \in \mathcal{S}ym^+$ defined as

$$\mathbf{C} = \mathbf{R}\mathbf{C}'\mathbf{R}^T, \tag{1.86}$$

which can be easily verified to satisfy

$$\mathbf{C}^2 = \mathbf{R}\mathbf{C}'\mathbf{R}^T\mathbf{R}\mathbf{C}'\mathbf{R}^T = \mathbf{R}\mathbf{C}'^2\mathbf{R}^T = \mathbf{R}\mathbf{S}'\mathbf{R}^T = \mathbf{S}.$$

Note that the definition is constructive: Given $\mathbf{S}$, one must first find its eigenvalues and eigenvectors, then calculate $\mathbf{R}$, and finally, using (1.85), compute $\mathbf{C}'$ and then from (1.86) $\mathbf{C}$.

1.20 Polar Theorem

Of fundamental importance in continuum mechanics is the following polar theorem:

Theorem 1.20.1 (Polar Decomposition Theorem) *Let* $\mathbf{F}$ *be an operator with* $\det \mathbf{F} > 0$. *There always exist unique operators* $\mathbf{R} \in \mathcal{R}ot$, $\mathbf{U} \in \mathcal{S}ym^+$, *and* $\mathbf{V} \in \mathcal{S}ym^+$ *such that*

$$\mathbf{F} = \mathbf{R}\mathbf{U} = \mathbf{V}\mathbf{R}. \tag{1.87}$$

Proof As we have seen before in (1.83), (1.84) the operator $\mathbf{S} = \mathbf{F}^T\mathbf{F} \in \mathcal{S}ym^+$ and, thus, using the procedure described above, we can define the square root operator of $\mathbf{S}$ denoted by $\mathbf{U}$:

$$\mathbf{U}^2 = \mathbf{F}^T\mathbf{F}.$$

To prove the first identity of (1.87), we need to verify that $\mathbf{R} = \mathbf{F}\mathbf{U}^{-1}$ is a rotation operator. Indeed,

$$\mathbf{R}^T\mathbf{R} = \left(\mathbf{F}\mathbf{U}^{-1}\right)^T\mathbf{F}\mathbf{U}^{-1} = \mathbf{U}^{-1}\mathbf{F}^T\mathbf{F}\mathbf{U}^{-1} = \mathbf{U}^{-1}\mathbf{U}^2\mathbf{U}^{-1} = \mathbf{I},$$

$$\det\mathbf{R} = \det\left(\mathbf{F}\mathbf{U}^{-1}\right) = \frac{\det\mathbf{F}}{|\det\mathbf{F}|} = 1.$$

Let $\mathbf{V} \in \mathcal{S}ym^+$ be the operator obtained by the similarity transformation of $\mathbf{U}$ with respect to $\mathbf{R}$:

$$\mathbf{V} = \mathbf{R}\mathbf{U}\mathbf{R}^T, \quad \mathbf{U} = \mathbf{R}^T\mathbf{V}\mathbf{R}. \tag{1.88}$$

Inserting $\mathbf{U}$ given by (1.88) into the first identity of (1.87), we obtain the second identity, and the theorem is proven. □

Note that the theorem is constructive, so starting from $\mathbf{F}$, it is possible to obtain $\mathbf{U}$ first, then $\mathbf{R}$, and finally $\mathbf{V}$.

Remark 5 It is easy to prove that if $\mathbf{U}$ have eigenvalue λ and eigenvector $\mathbf{d}$, then $\mathbf{V}$ have the same eigenvalue λ and eigenvector $\mathbf{Rd}$. In fact, if

$$\mathbf{U}\,\mathbf{d} = \lambda\,\mathbf{d}, \tag{1.89}$$

then by multiplying both sides by $\mathbf{R}$ and taking into account (1.87) follows:

$$\mathbf{V}\,\mathbf{Rd} = \lambda\,\mathbf{Rd}. \tag{1.90}$$

The physical meaning of the polar decomposition theorem will be evident in the next chapter, but it is already clear if we consider the action of $\mathbf{F}$ on an eigenvector $\mathbf{d}$ of $\mathbf{U}$ (see Fig. 1.9a). In fact, taking into account (1.89), we can first see a deformation

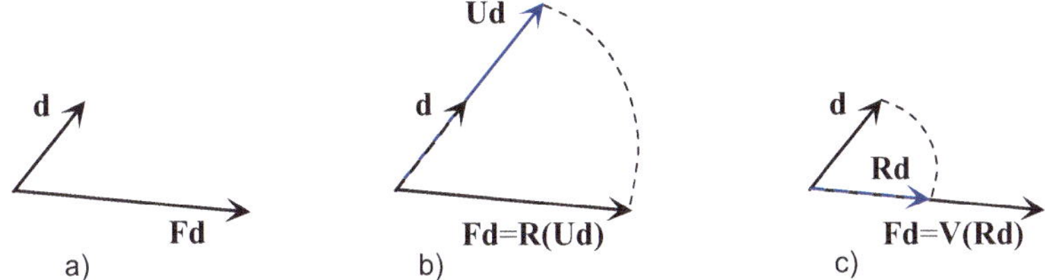

Fig. 1.9 Polar theorem

of **d** obtained by the action of **U** and then, from the first of (1.87), a rotation (see Fig. 1.9b). Alternatively, taking into account the second of (1.87), we can first see the action of the rotation by **R** on **d** and then, by (1.90), a deformation by the action of **V** (see Fig. 1.9c).

Chapter 2
Deformation of a Continuum

Abstract In this chapter, we will discuss the elements that characterize the deformation of a continuous medium.

2.1 Configuration of a Continuum

We consider a deformable continuous body in a configuration filled with material points of a regular region of the Euclidean space. Let C^* be the initial configuration of the continuum at time $t = 0$, and let C be the configuration at time t. We can represent a point P^* in C^* and its corresponding point P in C using position vectors as follows (see Fig. 2.1):

$$\mathbf{X} = OP^* \equiv (X_1, X_2, X_3), \quad \mathbf{x} = OP \equiv (x_1, x_2, x_3). \tag{2.1}$$

At any given time t, the vector function

$$\mathbf{x} \equiv \mathbf{x}(\mathbf{X}, t) \quad \Longleftrightarrow \quad x_i \equiv x_i(X_1, X_2, X_3, t), \quad i = 1, 2, 3, \tag{2.2}$$

provides a mapping between C^* and C. This mapping allows us to determine the position of any point P in C if we know its corresponding point P^* in C^*. We assume that the function (2.2) is at least twice continuously differentiable.

2.2 Deformation Gradient Operator

The *deformation gradient operator* $\mathbf{F}$ is defined as follows:

$$\mathbf{F} = \frac{\partial \mathbf{x}}{\partial \mathbf{X}} \equiv \|F_{iA}\|, \quad F_{iA} = \frac{\partial x_i}{\partial X_A}, \tag{2.3}$$

© The Author(s), under exclusive license to Springer Nature Switzerland AG 2024
T. Ruggeri, *Introduction to the Thermomechanics of Continua and Hyperbolic Systems*, La Matematica per il 3+2 167,
https://doi.org/10.1007/978-3-031-69951-1_2

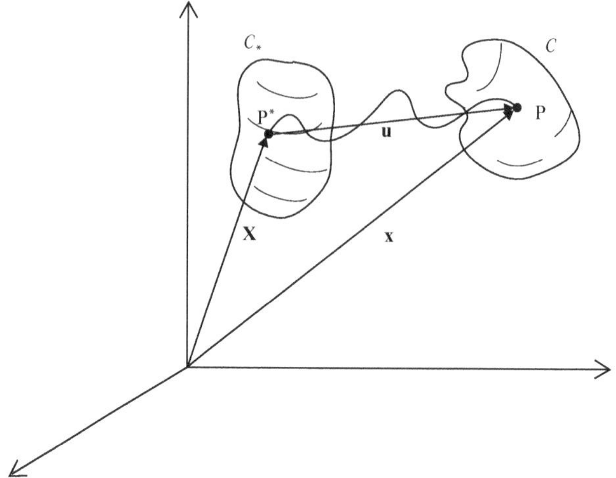

Fig. 2.1 Configuration of a continuum

where i and A range from 1 to 3. The components of $\mathbf{F}$ form the Jacobian matrix associated with the transformation (2.2).

To ensure that there is no overlap between two particles, we need to require at least local invertibility of the transformation (2.2). This implies that

$$J = \det(\mathbf{F}) \neq 0 \qquad \text{for all } \mathbf{X} \text{ and } t \geq 0.$$

Now at $t = 0$, $\mathbf{x} = \mathbf{X}$, then $\mathbf{F}|_{t=0} = \mathbf{I}$, and therefore $J|_{t=0} = 1$. In order to maintain continuity, $J(t)$ should remain positive for all times. If there is a time t^* such that $J(t^*) < 0$, then there would exist a time $\tilde{t} \in [0, t^*]$ for which $J(\tilde{t}) = 0$. Henceforth, we assume that in every configuration

$$J(\mathbf{X}, t) > 0.$$

In this case, the inverse transformation of (2.2) exists

$$\mathbf{X} \equiv \mathbf{X}(\mathbf{x}, t). \tag{2.4}$$

We can define the displacement vector as

$$\mathbf{u} = \mathbf{x} - \mathbf{X}, \tag{2.5}$$

which represents the displacement of a point from its initial position P^* to its current position P (see Fig. 2.1).

The *displacement gradient* operator, denoted as Grad $\mathbf{u}$, is defined as

$$\text{Grad}\,\mathbf{u} = \frac{\partial \mathbf{u}}{\partial \mathbf{X}} \equiv \left\| \frac{\partial u_i}{\partial X_A} \right\| . \tag{2.6}$$

From (2.5), we can derive a relationship between the deformation gradient and the displacement gradient operators:

$$\text{Grad}\,\mathbf{u} = \mathbf{F} - \mathbf{I}, \tag{2.7}$$

where $\mathbf{I}$ is the identity matrix.

The deformation gradient operator $\mathbf{F}$ and the displacement gradient operator Grad $\mathbf{u}$ provide important information about the deformation of the continuum. They describe the stretching, compression, and shearing of material elements within the continuum, and they play a crucial role in the analysis of the mechanical behavior of materials and structures.

2.3 Deformation Operators

Let $d\mathbf{X} = dP^* \equiv (dX_1, dX_2, dX_3)$ be an infinitesimal vector starting from P^* and having a point in the body very close to P^* as its second endpoint, and let $d\mathbf{x} = dP \equiv (dx_1, dx_2, dx_3)$ be its image vector in C (Fig. 2.2). From Eq. $(2.2)_2$, we immediately have

$$dx_i = \frac{\partial x_i}{\partial X_A} dX_A, \quad \text{or} \quad d\mathbf{x} = \mathbf{F} d\mathbf{X}. \tag{2.8}$$

Therefore, the operator $\mathbf{F}$ allows us to transform an infinitesimal vector in C^* to its corresponding vector in C.

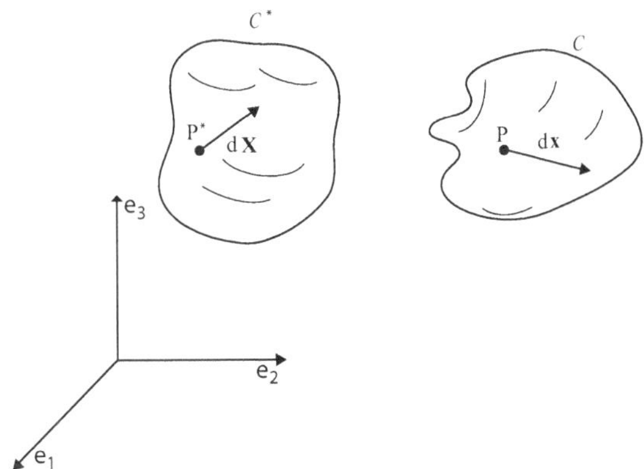

Fig. 2.2 Deformation

To measure the difference between the magnitudes of dP and dP^*, and thus the deviation from a rigid transformation, we calculate the squared norm of $d\mathbf{x}$ taking into account (1.7):

$$|d\mathbf{x}|^2 = d\mathbf{x} \cdot d\mathbf{x} = \mathbf{F}\,d\mathbf{X} \cdot \mathbf{F}\,d\mathbf{X} = \mathbf{F}^T\mathbf{F}\,d\mathbf{X} \cdot d\mathbf{X}. \tag{2.9}$$

Introducing the operator

$$\mathbf{C} = \mathbf{F}^T\mathbf{F}\,, \qquad C_{AB} = \frac{\partial x_i}{\partial X_A}\frac{\partial x_i}{\partial X_B}, \tag{2.10}$$

Eq. (2.9) becomes

$$|d\mathbf{x}|^2 = \mathbf{C}\,d\mathbf{X} \cdot d\mathbf{X}. \tag{2.11}$$

Note that, as mentioned in (1.83) and (1.84), $\mathbf{C} \in \mathcal{Sym}^+$, meaning that the operator $\mathbf{C}$ is symmetric and positive definite. Setting

$$\mathbf{E} = \frac{1}{2}(\mathbf{C} - \mathbf{I})\,, \quad E_{AB} = \frac{1}{2}\left(\frac{\partial x_i}{\partial X_A}\frac{\partial x_i}{\partial X_B} - \delta_{AB}\right), \tag{2.12}$$

Eq. (2.11) can also be written as

$$|d\mathbf{x}|^2 - |d\mathbf{X}|^2 = 2\mathbf{E}\,d\mathbf{X} \cdot d\mathbf{X}. \tag{2.13}$$

Therefore, in the case of rigid transformations, $\mathbf{E} = \mathbf{0}$ and $\mathbf{C} = \mathbf{I}$, which is why the operator $\mathbf{E}$ measures the deformation undergone by the body and is called the *Green–St.Venant deformation operator (or tensor)*. From (2.12), we have $\mathbf{E} \in \mathcal{Sym}$, but in general, it is not a positive definite operator.

The operator $\mathbf{C}$ is called the *right Cauchy–Green deformation operator (or tensor)*. However, note that in a rigid transformation, $\mathbf{C}$ is not zero but is equal to the identity operator.

Recall that the polar decomposition theorem (1.87) states that there always exist a rotation matrix $\mathbf{R}$ and an operator $\mathbf{U} \in \mathcal{Sym}^+$ such that:

$$\mathbf{F} = \mathbf{R}\mathbf{U}$$

with

$$\mathbf{U}^2 = \mathbf{F}^T\mathbf{F} = \mathbf{C}.$$

It follows that $\mathbf{U}$ is the square root operator of the Cauchy–Green deformation operator $\mathbf{C}$ (see Sect. 1.19.2):

$$\mathbf{U} = \sqrt{\mathbf{C}}. \tag{2.14}$$

Taking into account (2.7), it is possible to express the deformation tensors in terms of the displacement gradient rather than the deformation gradient:

$$\mathbf{C} = \mathbf{I} + \text{Grad}\mathbf{u} + (\text{Grad}\,\mathbf{u})^T + (\text{Grad}\,\mathbf{u})^T\,\text{Grad}\,\mathbf{u}$$

(CAUCHY-GREEN RIGHT);

$$C_{AB} = \delta_{AB} + \frac{\partial u_A}{\partial X_B} + \frac{\partial u_B}{\partial X_A} + \frac{\partial u_i}{\partial X_A}\frac{\partial u_i}{\partial X_B}$$

(2.15)

$$\mathbf{E} = \frac{1}{2}\left(\text{Grad}\mathbf{u} + (\text{Grad}\mathbf{u})^T + (\text{Grad}\,\mathbf{u})^T\,\text{Grad}\mathbf{u}\right)$$

(GREEN-SAINT VENANT);

$$E_{AB} = \frac{1}{2}\left(\frac{\partial u_A}{\partial X_B} + \frac{\partial u_B}{\partial X_A} + \frac{\partial u_C}{\partial X_A}\frac{\partial u_C}{\partial X_B}\right).$$

Therefore, given the displacement gradient Grad **u**, it is possible to determine the deformation tensors **C** (or **U**) and **E** and hence analyze the deformation and strain experienced by the body.

2.4 Inverse Deformation Operator

If we consider the inverse deformation (2.4), we can write

$$d\mathbf{X} = \mathbf{F}^{-1}d\mathbf{x},$$

which implies

$$|d\mathbf{X}|^2 = \left(\mathbf{F}^{-1}\right)^T \mathbf{F}^{-1}d\mathbf{x}\cdot d\mathbf{x}.$$

Defining:

$$\mathbf{c} = (\mathbf{F}^{-1})^T\mathbf{F}^{-1} \quad \text{and} \quad \mathbf{e} = \frac{1}{2}\left(\mathbf{I} - \mathbf{c}\right),$$

we obtain

$$|d\mathbf{X}|^2 = \mathbf{c}\,d\mathbf{x}\cdot d\mathbf{x} = |d\mathbf{x}|^2 - 2\mathbf{e}\,d\mathbf{x}\cdot d\mathbf{x},$$

i.e.,

$$|d\mathbf{x}|^2 - |d\mathbf{X}|^2 = 2\mathbf{e}\,d\mathbf{x}\cdot d\mathbf{x},$$

which corresponds to the form similar to Eq. (2.13).

The operators $\mathbf{c}$ and $\mathbf{e}$ are called the *Cauchy–Green* and *Green–St. Venant* operators of the inverse deformation, respectively. According to the polar decomposition theorem, we have

$$\mathbf{F} = \mathbf{V}\mathbf{R},$$

where $\mathbf{V} \in \mathcal{S}ym^{+}$ is the dilatation:

$$\mathbf{F}\mathbf{F}^{T} = \mathbf{V}^{2}.$$

It follows that

$$\left(\mathbf{V}^{2}\right)^{-1} = (\mathbf{F}^{T})^{-1}\mathbf{F}^{-1} = \mathbf{c},$$

and therefore,

$$\mathbf{V} = \sqrt{\mathbf{c}^{-1}}. \tag{2.16}$$

Finally, it should be noted that $\mathbf{c}^{-1}$ is not the same as $\mathbf{C}$, but it is related to it through the similarity transformation induced by the operator $\mathbf{R}$ in the polar decomposition theorem. In fact, from Eq. (1.88), it can be easily verified that

$$\mathbf{C} = \mathbf{R}^{T}\mathbf{c}^{-1}\mathbf{R}.$$

The tensor

$$\mathbf{B} = \mathbf{c}^{-1} = \mathbf{V}^{2} = \mathbf{F}\mathbf{F}^{T}$$

is called the *left Cauchy–Green deformation tensor*, while $\mathbf{e}$ is called the *Almansi strain tensor*. The relation (2.16) can be rewritten as

$$\mathbf{V} = \sqrt{\mathbf{B}}, \tag{2.17}$$

that is, the counterpart of (2.14). The positive eigenvalues $(\lambda_1, \lambda_2, \lambda_3)$ of $\mathbf{U}$ and $\mathbf{V}$ are the same (see Remark 5) and are called *principal stretch*. From (2.14) and (2.17) also $\mathbf{C}$ and $\mathbf{B}$ have the same eigenvalues that are $(\lambda_1^2, \lambda_2^2, \lambda_3^2)$.

2.5 Linear Dilatation Coefficient

Let us consider (see Fig. 2.3) a vector $d\mathbf{X}$ in the direction defined by the unit vector $\mathbf{u}_*$, that is,

$$d\mathbf{X} = dl^{*}\,\mathbf{u}_{*},$$

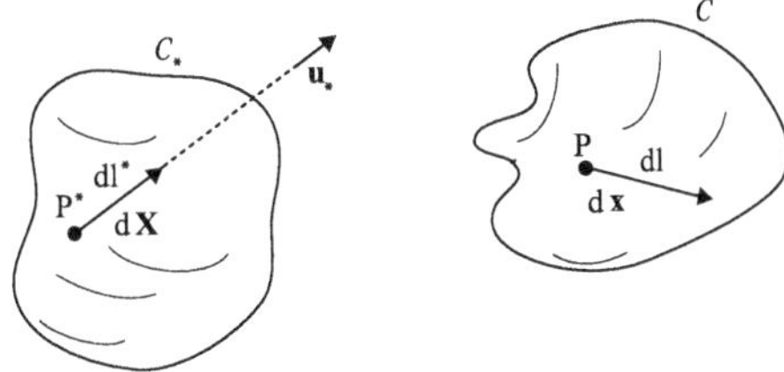

Fig. 2.3 Linear dilatation

where $dl^* = \|d\mathbf{x}_*\|$, and we will call the *linear dilatation coefficient* in the direction $\mathbf{u}_*$:

$$\delta_{u^*} = \frac{dl - dl^*}{dl^*}, \tag{2.18}$$

where $dl = \|d\mathbf{x}\|$. From (2.11) it immediately follows

$$dl = dl^* \sqrt{\mathbf{C}\,\mathbf{u}_* \cdot \mathbf{u}_*}, \tag{2.19}$$

and therefore substituting in (2.18), we obtain

$$\delta_{u^*} = \sqrt{\mathbf{C}\,\mathbf{u}_* \cdot \mathbf{u}_*} - 1. \tag{2.20}$$

As it can be seen, the linear dilatation coefficient depends on the operator $\mathbf{C}$ and also on the *initial fiber* identified by the unit vector $\mathbf{u}_*$.

Equation (2.20) can be written in terms of $\mathbf{E}$ as

$$\delta_{u^*} = \sqrt{1 + 2\mathbf{E}\,\mathbf{u}_* \cdot \mathbf{u}_*} - 1. \tag{2.21}$$

In particular, *the linear dilatation coefficients in the directions of the coordinate axes* are given by

$$\begin{aligned}
\delta_1 &= \sqrt{C_{11}} - 1 = \sqrt{1 + 2E_{11}} - 1, \\
\delta_2 &= \sqrt{C_{22}} - 1 = \sqrt{1 + 2E_{22}} - 1, \\
\delta_3 &= \sqrt{C_{33}} - 1 = \sqrt{1 + 2E_{33}} - 1.
\end{aligned} \tag{2.22}$$

Therefore, the diagonal elements of $\mathbf{C}$ and $\mathbf{E}$ are responsible for the linear dilatation in the directions $\mathbf{e}_1$, $\mathbf{e}_2$, and $\mathbf{e}_3$, respectively.

2.6 Shear

Let $d\mathbf{X}$ and $\delta\mathbf{X}$ be two infinitesimal vectors in C^* , taken in the directions of the unit vectors $\mathbf{u}_*$ and $\mathbf{v}_*$, and let $d\mathbf{x}$ and $\delta\mathbf{x}$ be the corresponding vectors in C. We denote the angle between $d\mathbf{x}$ and $\delta\mathbf{x}$ as $\theta_{(\mathbf{u}^*,\mathbf{v}^*)}$. We have

$$d\mathbf{x} \cdot \delta\mathbf{x} = \mathbf{F}d\mathbf{X} \cdot \mathbf{F}\delta\mathbf{X} = \mathbf{C}d\mathbf{X} \cdot \delta\mathbf{X}. \tag{2.23}$$

Denoting the magnitudes of $d\mathbf{X}$, $\delta\mathbf{X}$, $d\mathbf{x}$, and $\delta\mathbf{x}$ as dl^*, δl^*, dl, and δl, respectively, we can rewrite (2.23) as (see Fig. 2.4):

$$\cos\theta_{(\mathbf{u}^*,\mathbf{v}^*)} = \frac{dl^*}{dl}\frac{\delta l^*}{\delta l}\mathbf{C}\mathbf{u}_* \cdot \mathbf{v}_*.$$

Taking into account (2.19) and (2.20), we obtain

$$\cos\theta_{(\mathbf{u}^*,\mathbf{v}^*)} = \frac{\mathbf{C}\mathbf{u}_* \cdot \mathbf{v}_*}{(1+\delta_{u^*})(1+\delta_{v^*})}$$

or

$$\cos\theta_{(\mathbf{u}^*,\mathbf{v}^*)} = \frac{\mathbf{C}\mathbf{u}_* \cdot \mathbf{v}_*}{\sqrt{\mathbf{C}\mathbf{u}_* \cdot \mathbf{u}_*}\sqrt{\mathbf{C}\mathbf{v}_* \cdot \mathbf{v}_*}}. \tag{2.24}$$

In the case where $\mathbf{u}_*$ and $\mathbf{v}_*$ coincide with the directions of two of the reference axes, we obtain

$$\begin{aligned} \cos\theta_{12} &= \frac{C_{12}}{\sqrt{C_{11}}\sqrt{C_{22}}} = \frac{2E_{12}}{\sqrt{1+2E_{11}}\sqrt{1+2E_{22}}} \\ \cos\theta_{13} &= \frac{C_{13}}{\sqrt{C_{11}}\sqrt{C_{33}}} = \frac{2E_{13}}{\sqrt{1+2E_{11}}\sqrt{1+2E_{33}}} \\ \cos\theta_{23} &= \frac{C_{23}}{\sqrt{C_{22}}\sqrt{C_{33}}} = \frac{2E_{23}}{\sqrt{1+2E_{22}}\sqrt{1+2E_{33}}}. \end{aligned} \tag{2.25}$$

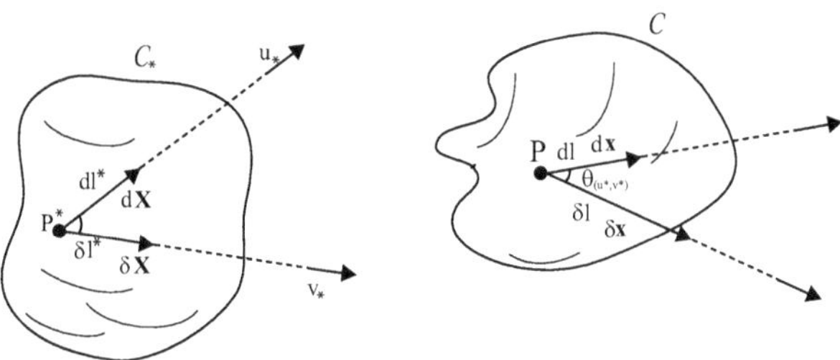

Fig. 2.4 Shear

The quantities

$$\tau_{ij} = \theta_{ij} - \frac{\pi}{2} \tag{2.26}$$

measure the difference between the angles after deformation and the initial angles and are called *shears*. From (2.26), we have

$$\sin \tau_{ij} = -\cos \theta_{ij}. \tag{2.27}$$

2.7 Surface Dilatation Coefficient

Let us denote by $d\Sigma$ and $\mathbf{n}$, respectively, the area of the parallelogram determined by the vectors $d\mathbf{x}$ and $\delta\mathbf{x}$ and the unit vector normal to the plane of the vectors in such a way that it holds (see Fig. 2.5):

$$d\mathbf{x} \times \delta\mathbf{x} = d\Sigma\, \mathbf{n}\,.$$

Similarly, in C^*, we have

$$d\mathbf{X} \times \delta\mathbf{X} = d\Sigma^*\, \mathbf{n}^*.$$

From properties (1.42), we immediately have

$$d\mathbf{x} \times \delta\mathbf{x} = \mathbf{F}d\mathbf{X} \times \mathbf{F}\delta\mathbf{X} = \mathbf{F}^C\,(d\mathbf{X} \times \delta\mathbf{X})$$

and therefore

$$d\Sigma\, \mathbf{n} = d\Sigma^*\, \mathbf{F}^C \mathbf{n}^*. \tag{2.28}$$

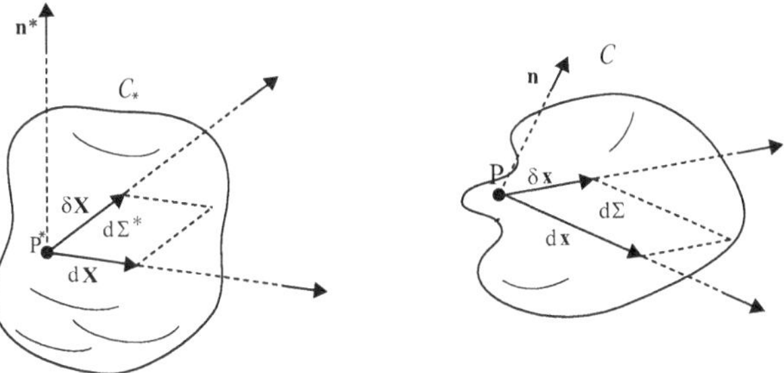

Fig. 2.5 Surface dilatation

Taking the squares of the vectors, we have

$$(d\Sigma)^2 = \left(d\Sigma^*\right)^2 \mathbf{F}^C\mathbf{n}^* \cdot \mathbf{F}^C\mathbf{n}^* = \left(d\Sigma^*\right)^2 \mathbf{F}^{CT}\mathbf{F}^C\, \mathbf{n}^* \cdot \mathbf{n}^*, \tag{2.29}$$

but from (1.19) and (2.10)

$$\mathbf{F}^{CT}\mathbf{F}^C = J^2\mathbf{F}^{-1}\mathbf{F}^{-1T} = J^2\left(\mathbf{F}^T\mathbf{F}\right)^{-1} = J^2\mathbf{C}^{-1},$$

and thus inserting into (2.29)

$$d\Sigma = d\Sigma^* J\sqrt{\mathbf{C}^{-1}\mathbf{n}^* \cdot \mathbf{n}^*}. \tag{2.30}$$

We call the surface dilatation coefficient

$$\delta_\Sigma = \frac{d\Sigma - d\Sigma^*}{d\Sigma^*},$$

and therefore we obtain

$$\delta_\Sigma = J\sqrt{\mathbf{C}^{-1}\mathbf{n}^* \cdot \mathbf{n}^*} - 1. \tag{2.31}$$

Note that δ_Σ depends not only on the deformation matrix $\mathbf{C}$ through its inverse but also on the normal vector $\mathbf{n}^*$ of the initial plane.

By substituting (2.30) into (2.28), we also have that the unit vector normal to the *deformed plane* is given by

$$\mathbf{n} = \frac{\mathbf{F}^C\mathbf{n}^*}{J\sqrt{\mathbf{C}^{-1}\mathbf{n}^* \cdot \mathbf{n}^*}}. \tag{2.32}$$

If the motion is rigid, we have $\mathbf{F} = \mathbf{R}$ and $\mathbf{C} = \mathbf{I}$, and therefore $\mathbf{n} = \mathbf{R}\mathbf{n}^*$ and $\delta_\Sigma = 0$.

2.8 Volume Dilatation Coefficient

Let us now consider three infinitesimal vectors $d\mathbf{x}$, $\delta\mathbf{x}$, and $\Delta\mathbf{x}$ starting from P in C and their corresponding image vectors be $d\mathbf{X}$, $\delta\mathbf{X}$, and $\Delta\mathbf{X}$ (see Fig. 2.6), and it immediately follows from (1.41):

$$dV = d\mathbf{x} \times \delta\mathbf{x} \cdot \Delta\mathbf{x} = J d\mathbf{X} \times \delta\mathbf{X} \cdot \Delta\mathbf{X} = J dV^*, \tag{2.33}$$

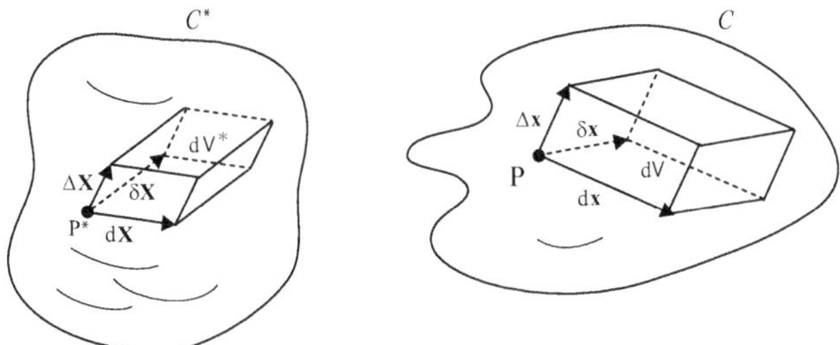

Fig. 2.6 Volume dilatation

where dV^* and dV denote the volumes determined by the three vectors in C^* and in C, respectively, yielding the volume dilatation coefficient:

$$\delta_V = \frac{dV - dV^*}{dV^*},$$

which is given by

$$\delta_V = J - 1. \tag{2.34}$$

2.9 Incompressible Bodies

Definition 2.9.1 A body is said to be incompressible if, even under deformation, its volume remains constant.

From (2.34), it immediately follows that the necessary and sufficient condition for the body to be incompressible is

$$J = 1. \tag{2.35}$$

2.10 Unimodular Deformation

Sometimes, it is better to use the first *unimodular* deformation tensor, denoted as

$$\bar{\mathbf{F}} = J^{-\frac{1}{3}}\mathbf{F}, \tag{2.36}$$

instead of the gradient of deformation $\mathbf{F}$. This tensor represents the part of $\mathbf{F}$ that accounts for shape changes without altering volume. In fact, due to property (1.10), we have $\det \bar{\mathbf{F}} = 1$.

This tensor induces the unimodular deformation tensors

$$\bar{\mathbf{B}} = \bar{\mathbf{F}}\bar{\mathbf{F}}^T = J^{-\frac{2}{3}}\mathbf{B}, \quad \bar{\mathbf{C}} = \bar{\mathbf{F}}^T\bar{\mathbf{F}} = J^{-\frac{2}{3}}\mathbf{C}, \tag{2.37}$$

which isolate the component of deformation involving shape changes without volume changes. These tensors are particularly important for elastic materials, as we will discuss later.

2.11 Homogeneous Deformation

Definition 2.11.1 A deformation is called homogeneous when $\mathbf{F}$ does not depend on $\mathbf{X}$.

We have

$$\frac{\partial \mathbf{x}}{\partial \mathbf{X}} = \mathbf{F}(t) \quad \Rightarrow \quad \mathbf{x}(\mathbf{X},t) = \mathbf{F}(t)\,\mathbf{X} + \mathbf{g}(t)$$

with $\mathbf{g}(0) = 0$ so that $\mathbf{x}(\mathbf{X},0) \equiv \mathbf{X}$ $(\mathbf{F}(0) \equiv \mathbf{I})$. Therefore, in a homogeneous deformation, the transformation that gives $\mathbf{x}$ from $\mathbf{X}$ is linear.

2.12 Small Deformations

In some engineering applications, the deformation is small. In such cases, the obtained formulas can be approximated and significantly simplified. Suppose that the components of the strain tensor $\mathbf{E}$ are so small that we can neglect the squares. In this case, by truncating the TAYLOR expansion at the first order, it can be seen from (2.21) and (2.22) that the linear dilatation coefficients take the approximate form:

$$\delta_{u^*} \simeq \mathbf{E}\mathbf{u}^* \cdot \mathbf{u}^*; \qquad \delta_1 \simeq E_{11}\,;\; \delta_2 \simeq E_{22}\,;\; \delta_3 \simeq E_{33}, \tag{2.38}$$

while from (2.27) and (2.25), the shear strains become

$$\tau_{ij} = -2E_{ij}. \tag{2.39}$$

Regarding the surface and volume dilatation coefficients, recalling that $\mathbf{C}\mathbf{C}^{-1} = \mathbf{I}$, for small deformations, we have

$$\mathbf{C}^{-1} \simeq \mathbf{I} - 2\mathbf{E},$$

and we also have

$$
\begin{aligned}
\det \mathbf{C} &= \det(\mathbf{I} + 2\mathbf{E}) \simeq 1 + 2(E_{11} + E_{22} + E_{33}) = 1 + 2 \, \mathrm{tr}\mathbf{E} \\
&\Longrightarrow \\
J &= \det \mathbf{F} = \sqrt{\det \mathbf{C}} \simeq 1 + \mathrm{tr}\mathbf{E};
\end{aligned}
\tag{2.40}
$$

thus, the surface dilatation coefficient (2.31) becomes

$$
\delta_\Sigma \simeq \mathrm{tr}\mathbf{E} - \mathbf{E}\mathbf{n}^* \cdot \mathbf{n}^*. \tag{2.41}
$$

In the case where $\mathbf{n}^*$ coincides respectively with $\mathbf{e}_1$, $\mathbf{e}_2$, and $\mathbf{e}_3$, we have

$$
\delta_{\Sigma_1} = E_{22} + E_{33}; \quad \delta_{\Sigma_2} = E_{11} + E_{33}; \quad \delta_{\Sigma_3} = E_{11} + E_{22}.
$$

The volume dilatation coefficient (2.34) becomes

$$
\delta_V \simeq \mathrm{tr}\,\mathbf{E}. \tag{2.42}
$$

Assuming small deformation is equivalent to assuming that the displacement gradient is small, in this case, the expressions (2.15) can be approximated as follows:

$$
\begin{aligned}
&\mathbf{C} = \mathbf{I} + \mathrm{Grad}\mathbf{u} + (\mathrm{Grad}\,\mathbf{u})^T, \qquad C_{AB} = \delta_{AB} + \frac{\partial u_A}{\partial X_B} + \frac{\partial u_B}{\partial X_A} \\
&\mathbf{E} = \frac{1}{2}\left(\mathrm{Grad}\mathbf{u} + (\mathrm{Grad}\mathbf{u})^T\right), \qquad E_{AB} = \frac{1}{2}\left(\frac{\partial u_A}{\partial X_B} + \frac{\partial u_B}{\partial X_A}\right).
\end{aligned}
\tag{2.43}
$$

From this, it can be inferred that *in the case of small deformations, the* GREEN–SAINT VENANT *strain tensor coincides with the symmetric part of the displacement gradient.*

Using (2.40) and (2.42) together with (2.43), we also have

$$
J \simeq 1 + \mathrm{Div}\,\mathbf{u}, \qquad \delta_V \simeq \mathrm{Div}\,\mathbf{u}, \qquad \left(\mathrm{Div}\,\mathbf{u} = \frac{\partial u_i}{\partial X_i}\right). \tag{2.44}
$$

Please note that these formulas are valid only for infinitesimal deformations and are commonly reported in texts on Structural Mechanics. For finite deformations, the expressions we derived earlier (2.15), (2.21), (2.24), (2.31), and (2.34) hold.

Chapter 3
Kinematics of a Continuum

Abstract In this chapter, we will provide an overview of the kinematic quantities of the motion of a continuum and define the velocity gradient operator and the vorticity and rate-of-deformation tensors.

3.1 Velocity and Acceleration

We define the velocity of a point P in the continuum as

$$\mathbf{v} = \frac{\partial \mathbf{x}(\mathbf{X}, t)}{\partial t},$$

where the velocity vector is a function of $\mathbf{X}$ and t:

$$\mathbf{v} \equiv \mathbf{v}(\mathbf{X}, t). \tag{3.1}$$

In this case, (3.1) represents the *Lagrangian velocity*. If we substitute $\mathbf{X}$ in (3.1) using (2.4), we can represent it as

$$\mathbf{v} \equiv \mathbf{v}(\mathbf{X}(\mathbf{x}, t), t) \equiv \mathbf{v}(\mathbf{x}, t),$$

which gives the velocity in the *Eulerian formulation*.

Similarly, we define the *Lagrangian acceleration* as

$$\mathbf{a}(\mathbf{X}, t) = \frac{\partial \mathbf{v}(\mathbf{X}, t)}{\partial t}$$

and the *Eulerian acceleration* as

$$\mathbf{a}(\mathbf{x}, t) = \frac{d}{dt}\mathbf{v}(\mathbf{x}(t), t).$$

© The Author(s), under exclusive license to Springer Nature Switzerland AG 2024

T. Ruggeri, *Introduction to the Thermomechanics of Continua and Hyperbolic Systems*, La Matematica per il 3+2 167,
https://doi.org/10.1007/978-3-031-69951-1_3

We can write the Eulerian acceleration as

$$a_i(\mathbf{x}, t) = \frac{\partial v_i}{\partial t} + \frac{\partial v_i}{\partial x_j}\frac{\partial x_j}{\partial t} = \frac{\partial v_i}{\partial t} + \frac{\partial v_i}{\partial x_j} v_j.$$

The operator

$$\frac{d}{dt} = \frac{\partial}{\partial t} + v_j \frac{\partial}{\partial x_j} \tag{3.2}$$

is called the *material time derivative*, which represents the time derivative of a quantity along the particle's path.

3.2 Velocity Gradient Operator

The velocity gradient operator is defined as

$$\nabla \mathbf{v} \equiv \left\| \frac{\partial v_i}{\partial x_j} \right\|.$$

We denote $\mathbf{D}$ and $\mathbf{W}$ as the symmetric and antisymmetric parts of $\nabla \mathbf{v}$, respectively:

$$\mathbf{D} = \frac{1}{2}\left(\nabla \mathbf{v} + (\nabla \mathbf{v})^T\right) \quad \Leftrightarrow \quad D_{ij} = \frac{1}{2}\left(\frac{\partial v_i}{\partial x_j} + \frac{\partial v_j}{\partial x_i}\right) \tag{3.3}$$

$$\mathbf{W} = \frac{1}{2}\left(\nabla \mathbf{v} - (\nabla \mathbf{v})^T\right) \quad \Leftrightarrow \quad W_{ij} = \frac{1}{2}\left(\frac{\partial v_i}{\partial x_j} - \frac{\partial v_j}{\partial x_i}\right). \tag{3.4}$$

Note that

$$\operatorname{tr} \mathbf{D} = \operatorname{tr} \nabla \mathbf{v} = \operatorname{div} \mathbf{v}.$$

From the properties of determinants $(1.80)_3$, we have

$$dJ = \mathbf{F}^C \cdot d\mathbf{F},$$

and therefore

$$\dot{J} = \mathbf{F}^C \cdot \dot{\mathbf{F}}. \tag{3.5}$$

On the other hand,

$$\dot{F}_{iA} = \frac{\partial}{\partial t}\left(\frac{\partial x_i}{\partial X_A}\right) = \frac{\partial}{\partial X_A}\left(\frac{\partial x_i}{\partial t}\right) = \frac{\partial v_i}{\partial X_A} = \frac{\partial v_i}{\partial x_k}\frac{\partial x_k}{\partial X_A},$$

which can be rewritten as

$$\dot{\mathbf{F}} = \nabla\mathbf{v}\,\mathbf{F}. \tag{3.6}$$

Thus, from (3.5),

$$\begin{aligned}\dot{J} &= \mathbf{F}^C \cdot \nabla\mathbf{v}\,\mathbf{F} = \nabla\mathbf{v}\,\mathbf{F}\cdot\mathbf{F}^C \\ &= \operatorname{tr}\left(\nabla\mathbf{v}\,\mathbf{F}\,\mathbf{F}^{CT}\right) = \operatorname{tr}\left(J\,\nabla\mathbf{v}\,\mathbf{F}\,\mathbf{F}^{-1}\right) = J\operatorname{tr}(\nabla\mathbf{v}) = J\operatorname{div}\mathbf{v}.\end{aligned}$$

Hence, we have the remarkable identity:

$$\dot{J} = J\operatorname{div}\mathbf{v}. \tag{3.7}$$

It immediately follows that if the body is incompressible, i.e., $J = 1$, the velocity field is *solenoidal*, i.e., $\operatorname{div}\mathbf{v} = 0$.

Now, we want to demonstrate that if the motion of the continuum is rigid, $\mathbf{D} = 0$ and $\nabla\mathbf{v} = \mathbf{W}$.

In a rigid motion, the distribution of velocities in the Eulerian view is given by

$$\mathbf{v}_P = \mathbf{v}_0 + \boldsymbol{\omega}\times OP, \quad \leftrightarrow \quad \mathbf{v}(\mathbf{x}) = \mathbf{v}(\mathbf{x}_0) + \boldsymbol{\omega}\times(\mathbf{x}-\mathbf{x}_0), \tag{3.8}$$

where $\mathbf{x}_0$ is a given point and $\boldsymbol{\omega}$ is the angular velocity independent of $\mathbf{x}$. From (3.8), we obtain

$$\nabla\mathbf{v} \equiv \begin{pmatrix} 0 & -\omega_3 & \omega_2 \\ \omega_3 & 0 & -\omega_1 \\ -\omega_2 & \omega_1 & 0 \end{pmatrix},$$

which is an antisymmetric matrix. Therefore,

$$\nabla\mathbf{v} = \mathbf{W} = \boldsymbol{\omega}\times\mathbf{v}, \tag{3.9}$$

where $\boldsymbol{\omega}$ is the dual vector of $\mathbf{W}$ (see 1.13.1). For this reason, the tensor $\mathbf{W}$ is called *vorticity* or *spin tensor*, while the tensor $\mathbf{D}$ is referred to as the *rate-of-deformation tensor*.

Chapter 4
Forces on a Continuum and Stress Tensor

Abstract This chapter presents the different types of forces acting on a continuum body. In particular, starting from the concept of *specific stress* and using CAUCHY'S theorem, it will be shown that the stress state is characterized by a symmetric second-order tensor.

4.1 Forces on a Continuum

Generally, we can have forces of different types acting on a continuum body:

1. ***External volume forces (or mass forces)***: At each internal point $P \in C$, consider a small volume element dV around P, and let dm be its mass. The forces that are proportional to the volume element and represented as

$$\mathbf{F} dV$$

are called *volume forces* (see Fig. 4.1). Note that $\mathbf{F}$ is a force per unit volume. Since $\rho = dm/dV$, it follows that

$$\mathbf{F} dV = \mathbf{b}\, dm$$

with

$$\mathbf{F} = \rho \mathbf{b}, \tag{4.1}$$

where $\mathbf{b}$ has the dimensions of acceleration (e.g., $\mathbf{b} = \mathbf{g}$ in the case of gravity). Forces of the form $\mathbf{b}\, dm$ are called *mass forces*.

© The Author(s), under exclusive license to Springer Nature Switzerland AG 2024
T. Ruggeri, *Introduction to the Thermomechanics of Continua and Hyperbolic Systems*, La Matematica per il 3+2 167,
https://doi.org/10.1007/978-3-031-69951-1_4

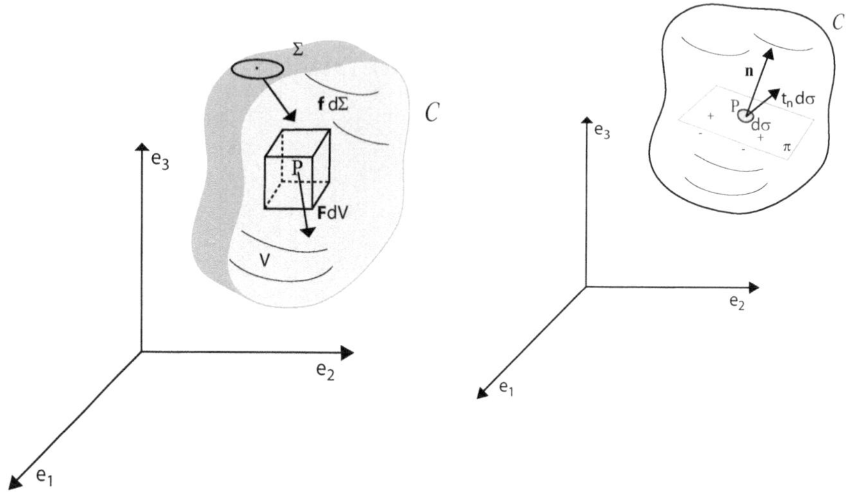

Fig. 4.1 Body, surface, and contact forces

2. ***External surface forces***: These forces act on the points $P \in \Sigma$ of the boundary and are represented as

$$\mathbf{f}\, d\Sigma,$$

where $\mathbf{f}$ has the dimensions of force per unit area (see Fig. 4.1).

3. ***Internal contact forces***: Consider a point $P \in C$ and a plane π passing through P with normal $\mathbf{n}$ (see Fig. 4.1). Taking a small surface element $d\sigma$ around P belonging to π, we define the positive side of $d\sigma$ as the side facing the direction of $\mathbf{n}$ and the negative side as the opposite side. The contact forces exerted by the molecules belonging to $d\sigma^+$ on those of $d\sigma^-$ will have a resultant represented as

$$\mathbf{t}_n\, d\sigma.$$

The index n indicates that $\mathbf{t}_n\, d\sigma$ depends not only on the point P but also, in general, on the choice of $\mathbf{n}$, i.e., the plane in which the section $d\sigma$ is considered. The dimensions of $\mathbf{t}_n$ are force per unit area, and $\mathbf{t}_n$ is called the *specific stress* in the direction $\mathbf{n}$.
By the principle of action and reaction, we have

$$\mathbf{t}_{-n} = -\mathbf{t}_n, \tag{4.2}$$

where $\mathbf{t}_{-n}$ represents the contact forces exerted by the molecules belonging to $d\sigma^-$ on those of $d\sigma^+$.

4.2 Cauchy's Theorem and Stress Tensor

For a given point $P \in C$, there are an infinite number of specific stresses corresponding to the infinite (∞^2) choices of $\mathbf{n}$.

However, there exists *Cauchy's theorem*, which allows us to obtain $\mathbf{t}_n$, as long as we know the specific stresses in the three normal directions corresponding to the chosen basis vectors $\mathbf{e}_1$, $\mathbf{e}_2$, and $\mathbf{e}_3$:

Theorem 4.2.1 (Cauchy's Theorem) *For any unit vector* $\mathbf{n} \equiv (n_1, n_2, n_3)$*, the following identity holds:*

$$\mathbf{t}_n = \mathbf{t}_1 n_1 + \mathbf{t}_2 n_2 + \mathbf{t}_3 n_3, \tag{4.3}$$

where $\mathbf{t}_n$ *represents the specific stress in the direction of* $\mathbf{n}$*, and* $\mathbf{t}_i$ *denotes the specific stresses in the directions* $\mathbf{n} \equiv \mathbf{e}_i$*, with* $i = 1, 2, 3$.

Proof Consider the body C in an equilibrium configuration. In this case, the first cardinal equation of statics states that the resultant of all external forces must be zero for any arbitrary portion of the continuum:

$$\mathbf{R}^{(e)} = 0.$$

Now, consider an internal portion of the continuum with volume ΔV and surface area $\Delta\sigma$. We have

$$\mathbf{R}^{(e)} = \int_{\Delta V} \mathbf{F} dV + \int_{\Delta\sigma} \mathbf{t}_n d\sigma = 0. \tag{4.4}$$

If we take a tetrahedron (as shown in Fig. 4.2) and denote the areas opposite to the vectors $\mathbf{e}_i$ as $\Delta\sigma_i$, the area of the oblique face as $\Delta\sigma$, and the volume of the

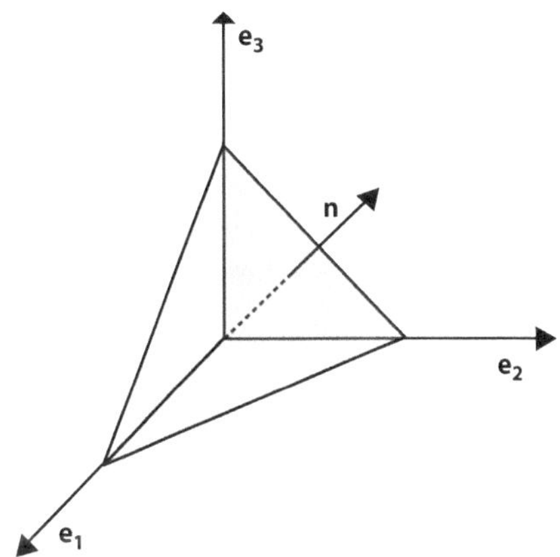

Fig. 4.2 Tetrahedron for Cauchy's theorem

tetrahedron as ΔV, then we can apply the mean value theorem to Eq. (4.4). This gives us

$$\mathbf{F}(\bar{P})\Delta V + \mathbf{t}_n(\tilde{P})\Delta\sigma - \mathbf{t}_1(\tilde{P}_1)\Delta\sigma_1 - \mathbf{t}_2(\tilde{P}_2)\Delta\sigma_2 - \mathbf{t}_3(\tilde{P}_3)\Delta\sigma_3 = 0, \tag{4.5}$$

where $\bar{P} \in \Delta V$, $\tilde{P} \in \Delta\sigma$, and $\tilde{P}_i \in \Delta\sigma_i$. For a tetrahedron, we know that $\Delta V = \Delta\sigma h/3$, and $\Delta\sigma_i = n_i \Delta\sigma$, where h is the height of the tetrahedron. Substituting these values in Eq. (4.5), we get

$$\mathbf{F}(\bar{P})\frac{h}{3} + \mathbf{t}_n(\tilde{P}) - \mathbf{t}_1(\tilde{P}_1)n_1 - \mathbf{t}_2(\tilde{P}_2)n_2 - \mathbf{t}_3(\tilde{P}_3)n_3 = 0. \tag{4.6}$$

Taking the limit as $h \to 0$, the first term vanish, and as $\tilde{P}$ and $\tilde{P}_i$ tend to P, we deduce Eq. (4.3). □

The significance of Cauchy's formula (4.3) is evident: *the knowledge of only the three components of each specific stress* $\mathbf{t}_i$ $(i = 1, 2, 3)$ *is sufficient to determine* $\mathbf{t}_n$.

Let us denote the components of the specific stresses as

$$\mathbf{t}_1 \equiv (t_{11}, t_{12}, t_{13}), \quad \mathbf{t}_2 \equiv (t_{21}, t_{22}, t_{23}), \quad \mathbf{t}_3 \equiv (t_{31}, t_{32}, t_{33}).$$

Equation (4.3) can then be written as

$$t_{ni} = t_{ji} n_j, \tag{4.7}$$

where t_{ni} denotes the i-th component of $\mathbf{t}_n$. By defining the matrix $\mathbf{t}$ with the specific stress components as its rows:

$$\mathbf{t} \equiv \begin{pmatrix} t_{11} & t_{12} & t_{13} \\ t_{21} & t_{22} & t_{23} \\ t_{32} & t_{32} & t_{33} \end{pmatrix},$$

Eq. (4.7) can be expressed as

$$\mathbf{t}_n = \mathbf{t}^T \mathbf{n}. \tag{4.8}$$

The matrix associated with the operator $\mathbf{t}$ is known as the *Cauchy stress matrix (or tensor)* and will be shown later to be symmetric, i.e., $\mathbf{t} \in \mathcal{Sym}$.

Remark 6 We have derived Cauchy's formula (4.3) assuming that the body is in equilibrium. However, the same formula holds even if the body is in motion. In fact, the duality from statics to dynamics, according to the principle of D'Alembert, is achieved by replacing active forces with the so-called *lost forces*:

$$\mathbf{F} \rightarrow \mathbf{F} - \rho\mathbf{a}. \tag{4.9}$$

As $h \rightarrow 0$, the validity of Cauchy's theorem in dynamic regime is obtained from Eq. (4.6) by substituting $\mathbf{F}$ with $\mathbf{F} - \rho\mathbf{a}$.

Chapter 5
Balance Laws in Continuum Mechanics

Abstract In this chapter, we will present the general laws that any continuous body must obey: *conservation of mass, balance of momentum, and balance of angular momentum.* Additionally, by using the *principle of virtual work*, we can calculate the work done by internal forces. Using the *transport theorem*, we will present the general mathematical structure of the balance laws in both the Eulerian and Lagrangian formulations. In the case of thermomechanical systems, the *balance of energy* is added to the aforementioned laws.

5.1 Conservation of Mass Law

The mass of a body or its part is described by the strictly positive density function ρ:

$$m = \int_C \rho \, dV.$$

The first conservation law in continuum mechanics states that every part of the body has the same mass during the evolution of the continuum. This law can be formulated in two different ways, depending on whether we are referring to a Lagrangian or Eulerian point of view.

5.1.1 Lagrangian Formulation

Referring to Fig. 5.1 and denoting ρ^* as the density in C^* and ρ as the density in C, we have

$$\int_{C^*} \rho^* dV^* = \int_C \rho dV, \tag{5.1}$$

© The Author(s), under exclusive license to Springer Nature Switzerland AG 2024

T. Ruggeri, *Introduction to the Thermomechanics of Continua and Hyperbolic Systems*, La Matematica per il 3+2 167,
https://doi.org/10.1007/978-3-031-69951-1_5

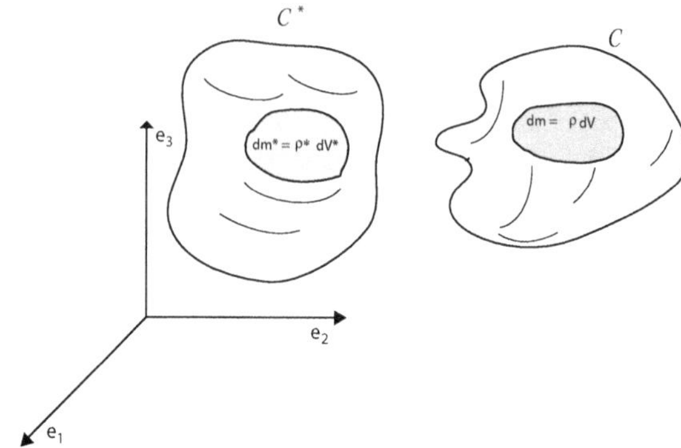

Fig. 5.1 Conservation of mass

but $dV = JdV^*$ (see (2.33)), and thus

$$\int_{C^*} (\rho J - \rho^*) dV^* = 0. \tag{5.2}$$

For (5.2) to hold for any domain, we have, pointwise,

$$\rho = \frac{\rho^*}{J}, \tag{5.3}$$

which expresses *the law of conservation of mass in Lagrangian variables* $(\mathbf{X}, t)$.

5.1.2 Eulerian Formulation

We have seen in (3.7) that

$$\frac{dJ}{dt} = J \operatorname{div} \mathbf{v},$$

and by differentiating (5.3), we obtain

$$\frac{d\rho}{dt} + \rho \operatorname{div} \mathbf{v} = 0. \tag{5.4}$$

Expanding the material derivative, we have

$$\frac{\partial \rho}{\partial t} + \frac{\partial \rho v_i}{\partial x_i} = 0, \tag{5.5}$$

which expresses *the law of conservation of mass in the Eulerian variables* $(\mathbf{x}, t)$.

5.2 Cardinal Equations

If a body in a configuration C is in equilibrium, then any internal generic part of the continuum will be also in equilibrium. Let us indicate the sub-body with ΔV and the regular frontier as $\Delta\sigma$. From Eq. (4.4), using CAUCHY'S THEOREM, we have

$$\int_{\Delta V} F_i \, dV + \int_{\Delta\sigma} t_{ji} n_j \, d\sigma = 0, \qquad i = 1, 2, 3. \tag{5.6}$$

We recall the GAUSS–GREEN THEOREM for which

$$\int_{\Delta\sigma} f n_k \, d\sigma = \int_{\Delta V} \frac{\partial f}{\partial x_k} \, dV, \tag{5.7}$$

where $\mathbf{n}$ is the outward unit normal vector to $\Delta\sigma$. By applying Eqs. (5.7) to (5.6), we get

$$\int_{\Delta V} \left(F_i + \frac{\partial t_{ji}}{\partial x_j} \right) dV = 0, \qquad i = 1, 2, 3. \tag{5.8}$$

For Eq. (5.8) to hold for any volume ΔV, we must have for regular solutions

$$F_i + \frac{\partial t_{ji}}{\partial x_j} = 0, \quad i = 1, 2, 3, \qquad \forall\, P \in V, \tag{5.9}$$

which represents the *first cardinal equation* of equilibrium.

The second cardinal equation states that the total moment of external forces with respect to any point Ω must be zero:

$$\mathbf{M}_{\Omega}^{(e)} = 0,$$

which, when we choose $\Omega \equiv O$, becomes

$$\int_{\Delta V} \mathbf{x} \times \mathbf{F} \, dV + \int_{\Delta\sigma} \mathbf{x} \times \mathbf{t}_n \, d\sigma = 0.$$

Taking the i-th component of the equation and using CAUCHY'S theorem and the LEVI-CIVITA symbol, we have

$$\int_{\Delta V} \varepsilon_{ilm} x_l F_m \, dV + \int_{\Delta\sigma} \varepsilon_{ilm} x_l t_{km} n_k \, d\sigma = 0, \qquad i = 1, 2, 3.$$

Applying the GAUSS–GREEN theorem to the second integral, we obtain

$$\int_{\Delta V} \varepsilon_{ilm} x_l F_m \, dV + \int_{\Delta V} \frac{\partial}{\partial x_k} (x_l t_{km}) \, \varepsilon_{ilm} \, dV = 0, \qquad i = 1, 2, 3.$$

Using the fact that $\frac{\partial x_l}{\partial x_k} = \delta_{lk}$, under regularity assumptions, we have

$$\varepsilon_{ilm} x_l \left(F_m + \frac{\partial t_{km}}{\partial x_k} \right) + \varepsilon_{ilm} t_{lm} = 0, \quad i = 1, 2, 3, \qquad \forall P \in V.$$

From Eq. (5.9), we see that the first term in the equation is identically zero, which implies

$$\varepsilon_{ilm} t_{lm} = 0, \qquad i = 1, 2, 3.$$

Taking $i = 1$, we get

$$\varepsilon_{1lm} \, t_{lm} = 0, \tag{5.10}$$

but in the sum on the indices l and m for the definition of the Levi-Civita symbol (1.1), the only nonzero values can be $l = 2$ and $m = 3$ or $l = 3$ and $m = 2$; then (5.10) becomes

$$\varepsilon_{123} \, t_{23} + \varepsilon_{132} \, t_{32} = 0, \qquad \leftrightarrow \qquad t_{23} - t_{23} = 0. \tag{5.11}$$

Proceeding in the same way choosing $i = 2, 3$, we obtain immediately

$$t_{23} = t_{32}, \quad t_{31} = t_{13}, \quad t_{12} = t_{21},$$

which means that $\mathbf{t}$ is necessarily symmetric:

$$\mathbf{t} = \mathbf{t}^T \in \mathcal{Sym}, \quad \leftrightarrow \quad t_{ij} = t_{ji}, \quad \forall (i, j) = 1, 2, 3. \tag{5.12}$$

In conclusion, the cardinal equations of statics reduce to Eq. (5.9) with $\mathbf{t} \in \mathcal{Sym}$. If we define the *divergence vector operator* as

$$\operatorname{div} \mathbf{t} \equiv (\operatorname{div} \mathbf{t})_i = \frac{\partial t_{ij}}{\partial x_j},$$

we can write Eq. (5.9) in vector form as

$$\mathbf{F} + \operatorname{div} \mathbf{t} = 0.$$

In the dynamic case, from Eq. (4.9), we immediately obtain the corresponding cardinal equation of dynamics:

$$\rho \mathbf{a} - \operatorname{div} \mathbf{t} = \mathbf{F}, \tag{5.13}$$

which represents the momentum theorem. Note that Eq. (5.13) is equivalent to

$$\rho \frac{d\mathbf{v}}{dt} - \operatorname{div} \mathbf{t} = \mathbf{F} \quad \Leftrightarrow \quad \rho \frac{dv_j}{dt} - \frac{\partial t_{ij}}{\partial x_i} = F_j, \quad (j = 1, 2, 3), \tag{5.14}$$

and by considering the mass balance Eqs. (5.5) and (4.1), it can also be written in the equivalent form:

$$\frac{\partial \rho v_j}{\partial t} + \frac{\partial}{\partial x_i} \left(\rho v_i v_j - t_{ij} \right) = \rho b_j, \qquad (j = 1, 2, 3). \tag{5.15}$$

5.2.1 Boundary Conditions

Consider a point on the surface where an external surface force $\mathbf{f} d\Sigma$ is applied, and let us consider the rectangular cylinder of length l and basis area $d\Sigma$, with radius dr. Let $\Delta\sigma'$ be the lateral surface of the cylinder of volume ΔV (see Fig. 5.2). In equilibrium conditions, Eq. (4.4) must hold

$$\int_{\Delta V} \mathbf{F} dV + \int_{\Delta \sigma'} \mathbf{t}_n d\sigma' + \mathbf{t}_n d\Sigma + \mathbf{f} d\Sigma = 0.$$

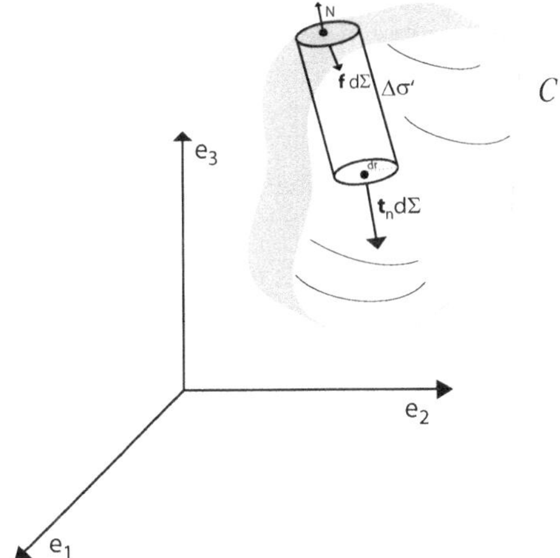

Fig. 5.2 Boundary conditions

Applying the mean value theorem,

$$\mathbf{F}(\bar{P})\, l\, d\Sigma + 2\pi\, l\, dr\, \mathbf{t}_n(\tilde{P}) + (\mathbf{f} + \mathbf{t}_n) d\Sigma = 0, \tag{5.16}$$

where $\bar{P} \in \Delta V$ and $\tilde{P} \in \Delta\sigma'$. Taking the limit as $l \to 0$ and denoting $\mathbf{N}$ as the outward unit normal vector to the surface (see Fig. 5.2), from (5.16) the following boundary conditions must hold for points $P \in \Sigma$:

$$\begin{aligned} &\mathbf{t}_N = \mathbf{t}_{-n} = -\mathbf{t}_n = \mathbf{f} \quad \Leftrightarrow \\ &\mathbf{t}_N = \mathbf{f} \quad \Leftrightarrow \quad t_{ij} N_j = f_i, \quad (i = 1, 2, 3), \quad \forall P \in \Sigma. \end{aligned} \tag{5.17}$$

5.3 Principle of Virtual Work

If we assume *ideal constraints*, we can obtain the same results as the equilibrium equations (5.9) and (5.12) by using the *principle of virtual work*. According to this principle, a necessary and sufficient condition for C to be an equilibrium configuration is that the work of active forces

$$\delta L^{(a)} \leq 0 \quad \forall \text{ virtual displacement } \delta P \text{ starting from } C. \tag{5.18}$$

In continuum mechanics, it is sometimes preferable to speak of the *principle of virtual power*, replacing the virtual displacement δP with the *virtual velocity* $\mathbf{v}$ (any velocity compatible with the constraints at the given instant).

The principle of virtual power is very useful because it not only allows us to recover (5.9) and (5.12) but also provides us with the expression for the power of the internal forces.

Let us assume bilateral constraints, so that the necessary and sufficient condition for a configuration C to be in equilibrium is that the power of the active forces (external and internal) is zero for every virtual velocity field starting from C:

$$P^{(a)} = P^{(e,a)} + P^{(i,a)} = 0. \tag{5.19}$$

Considering a volume element ΔV with boundary $\Delta\sigma$, we have

$$P^{(e,a)} = \int_{\Delta V} \mathbf{F} \cdot \mathbf{v} dV + \int_{\Delta\sigma} \mathbf{t}_n \cdot \mathbf{v} d\sigma.$$

Let us denote $\mathcal{P}^{(i)}$ as the power per unit volume of the active internal forces:

$$P^{(i,a)} = \int_{\Delta V} \mathcal{P}^{(i)} dV.$$

Substituting into (5.19), we obtain

$$\int_{\Delta V} F_i v_i dV + \int_{\Delta \sigma} t_{ij} v_i n_j d\sigma + \int_{\Delta V} \mathcal{P}^{(i)} dV = 0, \quad \forall \mathbf{v}.$$

Applying Gauss–Green's theorem, we obtain

$$\left(F_i + \frac{\partial t_{ij}}{\partial x_j} \right) v_i + t_{ij} \frac{\partial v_i}{\partial x_j} + \mathcal{P}^{(i)} = 0, \quad \forall \mathbf{v}. \tag{5.20}$$

To satisfy (5.20) for any virtual velocity, let us consider a translation rigid motion:

$$\mathbf{v} \equiv \mathbf{v}_\Omega = \text{constant}.$$

From Rational Mechanics, we recall that in the case of rigid motions, the power of any system of forces has a particular expression:

$$\mathcal{P} = \mathbf{R} \cdot \mathbf{v}_\Omega + \mathbf{M}_\Omega \cdot \boldsymbol{\omega},$$

where $\boldsymbol{\omega}$ is the angular velocity. Therefore, since the internal forces consist of zero-moment couples (by the action–reaction principle), we have $\mathbf{R}^{(i)} = 0$ and $\mathbf{M}_\Omega^{(i)} = 0$, and hence $\mathcal{P}^{(i)} = 0$ for any rigid motion. Thus, (5.20) becomes

$$\left(F_i + \frac{\partial t_{ij}}{\partial x_j} \right) v_{\Omega_i} = 0, \quad \forall \mathbf{v}_\Omega.$$

This implies (5.9). Then (5.20) reduces to

$$\mathbf{t} \cdot \nabla \mathbf{v} + \mathcal{P}^{(i)} = 0, \quad \forall \mathbf{v}. \tag{5.21}$$

Now, considering a general rigid motion (in particular, a rotational one)

$$\mathbf{v} = \mathbf{v}_\Omega + \boldsymbol{\omega} \times \mathbf{x}$$

and recalling that (see (3.9)) $\mathbf{D} = 0$, $\nabla \mathbf{v} = \mathbf{W} \in \mathcal{Skw}$, and $\mathcal{P}^{(i)} = 0$, we immediately have

$$\mathbf{t} \cdot \mathbf{W} = 0, \quad \forall\, \mathbf{W} \in \mathcal{Skw}.$$

Using Corollary 1 of Theorem 1.13.1 (see (1.47)), we can conclude that $\mathbf{t} \in \mathcal{Sym}$, and thus (5.12) is true.

Taking into account the symmetry of $\mathbf{t}$, (5.21) provides us with *the specific power of the internal forces*:

$$\mathcal{P}^{(i)} = -\mathbf{t} \cdot \mathbf{D}. \tag{5.22}$$

Thus, up to a sign, it is equal to the dot product between the stress tensor and the rate-of-deformation tensor.

When expressed explicitly in terms of components, (5.22) becomes

$$\mathcal{P}^{(i)} = -\frac{1}{2}t_{ij}\left(\frac{\partial v_i}{\partial x_j} + \frac{\partial v_j}{\partial x_i}\right). \tag{5.23}$$

5.4 General Balance Laws

Summarizing, we have presented the laws that every material body must obey in Eulerian variables:

$$\begin{cases} \dfrac{\partial \rho}{\partial t} + \dfrac{\partial \rho v_i}{\partial x_i} = 0 \quad \text{(conservation of mass)} \\ \\ \dfrac{\partial \rho v_j}{\partial t} + \dfrac{\partial}{\partial x_i}\left(\rho v_i v_j - t_{ij}\right) = \rho b_j \quad \text{(balance of momentum),} \end{cases} \tag{5.24}$$

where $t_{ij} = t_{ji}$. These equations represent a particular case of balance laws.

A law is called a *balance law for a given density* $\Psi(\mathbf{x}, t)$ if it can be represented as

$$\frac{d}{dt}\int_V \Psi dV = -\int_\Sigma \Phi_i n_i d\Sigma + \int_V f dV, \tag{5.25}$$

where Φ_i represents the *flux* and f represents any *sources.*

In essence, Eq. (5.25) states that the temporal variation of the total density

$$\frac{d}{dt}\int_V \Psi dV$$

must be equal to the normal component of the *flux* of a certain quantity through the surface

$$-\int_\Sigma \Phi_i n_i d\Sigma$$

plus any contributions from sources

$$\int_V f dV.$$

Depending on the physical meaning of Ψ, Φ_i, and f, we will have the corresponding balance laws. If there is no source term, the balance law is called a *conservation law.*

The first term in (5.25) involves the derivative of an integral whose integration domain V changes over time, so the differentiation operator cannot be moved inside the integral sign. To do so, we need to resort to the so-called *transport theorem*.

5.4.1 Transport Theorem

REYNOLD's *transport theorem* states

$$\frac{d}{dt}\int_V \Psi dV = \int_V \left(\frac{d\Psi}{dt} + \Psi \text{div}\mathbf{v}\right) dV. \tag{5.26}$$

The proof is straightforward by transforming the integral from the volume V to V^* and taking into account Eq. (3.7). We have

$$\frac{d}{dt}\int_V \Psi dV = \frac{d}{dt}\int_{V*} \Psi J dV^* = \int_{V*} \frac{d}{dt}(\Psi J)\, dV^*$$

$$= \int_{V*} \left(\frac{d\Psi}{dt} + \Psi \text{div}\mathbf{v}\right) J dV^* = \int_V \left(\frac{d\Psi}{dt} + \Psi \text{div}\mathbf{v}\right) dV.$$

Note that the transformation from V to V^* allows us to move the differentiation operator d/dt inside the integral sign since V^* is a constant domain. Then, using Eq. (3.7), we return from V^* to V.

Substituting Eq. (5.26) into the expression of the balance law and applying Gauss–Green's theorem, we immediately obtain

$$\int_V \left(\frac{d\Psi}{dt} + \Psi \text{div}\mathbf{v}\right) dV + \int_V \frac{\partial \Phi_i}{\partial x_i} dV = \int_V f dV. \tag{5.27}$$

Since Eq. (5.27) must hold for every domain V, if the appropriate regularity conditions are satisfied, we have

$$\frac{d\Psi}{dt} + \Psi \text{div}\mathbf{v} + \frac{\partial \Phi_i}{\partial x_i} = f, \tag{5.28}$$

which can be written locally as

$$\frac{\partial \Psi}{\partial t} + \frac{\partial}{\partial x_i}(\Psi v_i + \Phi_i) = f. \tag{5.29}$$

Hence, every balance law can be locally expressed in the form (5.29). Depending on the expressions assigned to Ψ, Φ_i, and f, we obtain different balance laws.

For example, by choosing

$$\Psi = \rho, \quad \Phi_i = 0, \quad f = 0,$$

Eq. (5.29) coincides with the conservation law of mass $(5.24)_1$. Note that there is no mass flow through the surface.

On the other hand, if we choose

$$\Psi = \rho v_j, \quad \Phi_i = -t_{ij}\,, \quad f = \rho b_j \qquad (j = 1, 2, 3)\,,$$

we obtain the three scalar equations of the balance law of momentum $(5.24)_2$.

5.4.2 Energy Balance Law

If we extend our attention from mechanics to thermomechanics, we encounter another important law: the *energy balance law*, which constitutes the *first law of thermodynamics*. In this case, the total energy density consists of two terms: the density of *kinetic energy* $\rho v^2/2$ and the density of *internal energy* ρe:

$$\Psi = \rho e + \frac{\rho v^2}{2},$$

while the flux $\Phi_i n_i$ consists of a flux of *mechanical energy* (*power of contact forces*) $-\mathbf{t}_n \cdot \mathbf{v}$ and the normal component of the *heat flux* $\mathbf{q} \cdot \mathbf{n}$; thus

$$\Phi_i = -t_{ij} v_j + q_i\,.$$

Finally, the source term is given by the *power of volume forces* $\mathbf{F} \cdot \mathbf{v}$ and any *heat sources* r (*radiation*):

$$f = \mathbf{F} \cdot \mathbf{v} + r = \rho \mathbf{b} \cdot \mathbf{v} + r.$$

From Eq. (5.29), we obtain

$$\frac{\partial}{\partial t}\left(\frac{1}{2}\rho v^2 + \rho e\right) + \frac{\partial}{\partial x_i}\left\{\left(\frac{1}{2}\rho v^2 + \rho e\right) v_i - t_{ij} v_j + q_i\right\} = \rho b_i v_i + r. \qquad (5.30)$$

In agreement with the first law of thermodynamics, if the system is isolated, i.e., $\mathbf{b} = 0$, $r = 0$, (5.30) represents the conservation of energy.

5.4.3 Thermomechanical Balance Laws in Eulerian Form

In the thermomechanical case, we have therefore three balance laws: the conservation law of mass, the balance of momentum, and the energy balance, which can be written in Eulerian variables as

$$\begin{cases} \dfrac{\partial \rho}{\partial t} + \dfrac{\partial \rho v_i}{\partial x_i} = 0 \\ \\ \dfrac{\partial \rho v_j}{\partial t} + \dfrac{\partial}{\partial x_i}\left(\rho v_i v_j - t_{ij}\right) = \rho b_j \qquad (j = 1, 2, 3) \\ \\ \dfrac{\partial}{\partial t}\left(\frac{1}{2}\rho v^2 + \rho e\right) + \dfrac{\partial}{\partial x_i}\left\{\left(\frac{1}{2}\rho v^2 + \rho e\right) v_i - t_{ij} v_j + q_i\right\} = \rho b_j v_j + r. \end{cases} \tag{5.31}$$

Sometimes it is convenient to represent the system (5.31) of five scalar equations as a single balance law in vectorial form:

$$\frac{\partial \mathbf{F}^0}{\partial t} + \frac{\partial \mathbf{F}^i}{\partial x^i} = \mathbf{f}, \tag{5.32}$$

where $\mathbf{F}^0$, $\mathbf{F}^i$ $(i = 1, 2, 3)$, and $\mathbf{f}$ are five-component vectors in this case:

$$\mathbf{F}^0 = \begin{pmatrix} \rho \\ \rho v_1 \\ \rho v_2 \\ \rho v_3 \\ \frac{1}{2}\rho v^2 + \rho e \end{pmatrix}, \mathbf{F}^i = \begin{pmatrix} \rho v_i \\ \rho v_i v_1 - t_{i1} \\ \rho v_i v_2 - t_{i2} \\ \rho v_i v_3 - t_{i3} \\ \left(\frac{1}{2}\rho v^2 + \rho e\right) v_i - t_{ij} v_j + q_i \end{pmatrix},$$

$$\mathbf{f} = \begin{pmatrix} 0 \\ \rho b_1 \\ \rho b_2 \\ \rho b_3 \\ \rho b_j v_j + r \end{pmatrix}. \tag{5.33}$$

Any system of N balance laws can be rewritten in the form (5.32) with $\mathbf{F}^0$, $\mathbf{F}^i$, and $\mathbf{f}$ as vectors in $\mathbb{R}^N$.

Taking inspiration from the relativistic context, it is convenient to consider time as the fourth coordinate $x^0 = t$ and introduce $\mathbf{F}^\alpha$ with $\alpha = 0, 1, 2, 3$. In this way, (5.32) becomes

$$\frac{\partial \mathbf{F}^\alpha}{\partial x^\alpha} = \mathbf{f}, \tag{5.34}$$

where now the sum over α runs from 0 to 3.

From (5.34), *we can clearly see the structure of the balance laws: The left-hand side of* (5.34) *represents the divergence in space-time of* $\mathbf{F}^\alpha$.

5.4.4 Galilean Invariance

The balance Eqs. (5.13) are invariant under Galilean transformations in accordance with the principle of relativity by GALILEO. In fact, considering the transformation

$$t' = t; \quad \mathbf{x}' = \mathbf{x} - \mathbf{c}t; \quad \mathbf{v}' = \mathbf{v} - \mathbf{c} \tag{5.35}$$

corresponding to two reference frames, one in uniform translation motion with a constant velocity $\mathbf{c}$ relative to the other, the laws remain unchanged.

An interesting problem is to determine the conditions under which a generic system of N balance Eqs. (5.32) is invariant under Galilean transformations (5.35).

It can be proven (see RUGGERI [11]) that the systems (5.32) satisfying the principle of Galilean relativity are those for which there exists a matrix $\mathbf{X}$ ($N \times N$) dependent only on velocity, satisfying

$$\begin{cases} \mathbf{F}^0(\mathbf{v}, \mathbf{w}) = \mathbf{X}(\mathbf{v})\mathbf{F}^0(\mathbf{0}, \mathbf{w}) \\ \mathbf{G}^i(\mathbf{v}, \mathbf{w}) = \mathbf{X}(\mathbf{v})\mathbf{G}^i(\mathbf{0}, \mathbf{w}) \\ \mathbf{f}(\mathbf{v}, \mathbf{w}) = \mathbf{X}(\mathbf{v})\mathbf{f}(\mathbf{0}, \mathbf{w}), \end{cases} \tag{5.36}$$

where

$$\mathbf{G}^i = \mathbf{F}^i - \mathbf{F}^0 v^i \tag{5.37}$$

and $\mathbf{w}$ represents the set of other physical variables that are assumed to be independent of the observer (objective quantities).

It can be shown that the matrix $\mathbf{X}(\mathbf{v})$ is an exponential matrix

$$\mathbf{X}(\mathbf{v}) = e^{\mathbf{A}^r v_r} = \mathbf{I} + \mathbf{A}^r v_r + \frac{1}{2}\mathbf{A}^r\mathbf{A}^s v_r v_s + \dots, \tag{5.38}$$

where $\mathbf{A}^r$ ($r = 1, 2, 3$) are three constant matrices that commute with each other

$$\mathbf{A}^r\mathbf{A}^s = \mathbf{A}^s\mathbf{A}^r, \qquad \forall\ r, s = 1, 2, 3. \tag{5.39}$$

In other words, Eq. (5.36) establishes that Galilean balance laws have a particular dependence on velocity. In physical cases, in which there exists an order in the balance laws, it can be shown that the matrices $\mathbf{A}^r$ are *nilpotent* of degree m, meaning that there exists an integer m such that

$$\mathbf{A}^{k_1}\mathbf{A}^{k_2}...\mathbf{A}^{k_m} = 0, \qquad \forall\, k_1,\, k_2,\, ...,\, k_m = 1,\, 2,\, 3, \tag{5.40}$$

and thus, from Eq. (5.38), it follows that the matrix $\mathbf{X}(\mathbf{v})$ is a polynomial matrix of degree $m - 1$.

Let us verify this result in the case of the system (5.31). From (5.33) and (5.37), we have

$$\mathbf{F}^0 \equiv \begin{pmatrix} \rho \\ \rho\, v_1 \\ \rho\, v_2 \\ \rho\, v_3 \\ \frac{1}{2}\rho v^2 + \rho\, e \end{pmatrix}, \quad \mathbf{G}^i \equiv \begin{pmatrix} 0 \\ -t^{i1} \\ -t^{i2} \\ -t^{i3} \\ -t^{ij} v_j + q_i \end{pmatrix}, \quad \mathbf{f} \equiv \begin{pmatrix} 0 \\ \rho\, b_1 \\ \rho\, b_2 \\ \rho\, b_3 \\ \rho\, b_j v_j + r \end{pmatrix}.$$

Using a hat to denote quantities evaluated at $\mathbf{v} = 0$, we have

$$\hat{\mathbf{F}}^0 \equiv \begin{pmatrix} \rho \\ 0 \\ 0 \\ 0 \\ \rho\, e \end{pmatrix}, \quad \hat{\mathbf{G}}^i \equiv \begin{pmatrix} 0 \\ -t^{i1} \\ -t^{i2} \\ -t^{i3} \\ q_i \end{pmatrix}, \quad \hat{\mathbf{f}} \equiv \begin{pmatrix} 0 \\ \rho\, b_1 \\ \rho\, b_2 \\ \rho\, b_3 \\ r \end{pmatrix}.$$

It is easy to verify that the matrix $\mathbf{X}(\mathbf{v})$ that satisfies all the equations (5.36) in this case is

$$\mathbf{X}(\mathbf{v}) \equiv \begin{pmatrix} 1 & 0 & 0 & 0 & 0 \\ v_1 & 1 & 0 & 0 & 0 \\ v_2 & 0 & 1 & 0 & 0 \\ v_3 & 0 & 0 & 1 & 0 \\ \frac{1}{2}v^2 & v_1 & v_2 & v_3 & 1 \end{pmatrix}.$$

The reader can verify that the $\mathbf{A}^r$ matrices are given by

$$\mathbf{A}^r = \left(\frac{\partial \mathbf{X}(\mathbf{v})}{\partial v_r}\right)_{\mathbf{v}=0} \equiv \begin{pmatrix} 0 & 0 & 0 & 0 & 0 \\ \delta^{1r} & 0 & 0 & 0 & 0 \\ \delta^{2r} & 0 & 0 & 0 & 0 \\ \delta^{3r} & 0 & 0 & 0 & 0 \\ 0 & \delta^{1r} & \delta^{2r} & \delta^{3r} & 0 \end{pmatrix},$$

which are nilpotent matrices of degree $m = 3$ for every r, and Eqs. (5.38)–(5.40) are satisfied.

5.4.5 Lagrangian Formulation of Balance Laws

In many cases, it is convenient to express the balance laws (5.31) in Lagrangian variables rather than Eulerian variables, by transforming the integrals in V and Σ into their Lagrangian counterparts V^* and Σ^*. Regarding the volume integrals, we already know how to transform them using (2.33):

$$dV = JdV^*, \tag{5.41}$$

while for the surface integrals, we need to recall the formula (2.28):

$$\mathbf{n}d\Sigma = \mathbf{F}^C\mathbf{n}^*d\Sigma^*. \tag{5.42}$$

Therefore, the balance law

$$\frac{d}{dt}\int_V \Psi dV = -\int_\Sigma \Phi_i n_i d\Sigma + \int_V f dV \tag{5.43}$$

can be rewritten as[1]

$$\frac{d}{dt}\int_{V^*} J\Psi dV^* = -\int_{\Sigma^*} \Phi_i F^C_{iA} n^*_A d\Sigma^* + \int_{V^*} fJdV^*. \tag{5.44}$$

Taking into account that V^* is constant, and therefore we can bring the derivative inside the first integral, and applying the GAUSS–GREEN theorem to the surface integral, we immediately obtain

$$\int_{V^*}\frac{\partial}{\partial t}(J\Psi)\, dV^* + \int_{V^*}\frac{\partial \Phi_i F^C_{iA}}{\partial X_A} dV^* = \int_{V^*} fJdV^*,$$

which, under regularity assumptions, becomes

$$\rho^*\frac{\partial}{\partial t}\left(\frac{\Psi}{\rho}\right) + \frac{\partial}{\partial X_A}\left(\Phi_i F^C_{iA}\right) = \frac{\rho^*}{\rho} f. \tag{5.45}$$

By introducing the quantities

$$\Psi^* = \frac{\rho^*}{\rho}\Psi, \quad \Phi^*_A = \Phi_i F^C_{iA}, \quad f^* = \frac{\rho^*}{\rho} f, \tag{5.46}$$

[1] To avoid complicating the notation, we will use the same symbols for Ψ, Φ_i, and f in (5.43) and (5.44); however, it should be noted that while in (5.43) these quantities are functions of $(\mathbf{x}, t)$, in (5.44) they are functions of $(\mathbf{X}, t)$ through the transformation (2.2).

the balance law in Lagrangian variables takes the form

$$\frac{\partial \Psi^*}{\partial t} + \frac{\partial \Phi_A^*}{\partial X_A} = f^*, \tag{5.47}$$

which is very similar to the Eulerian form (5.29) with the formal change from unprimed variables to primed variables and without the convective terms. *Therefore,* (5.47) *represents the Lagrangian formulation of a generic balance law.*

5.4.6 Balance Law of Momentum in Lagrangian Form and the First Piola–Kirchhoff Stress Tensor

In the case of the momentum balance,

$$\begin{gathered} \Psi = \rho v_j, \quad \Phi_i = -t_{ij}, \quad f = \rho b_j \\ \Longrightarrow \\ \Psi^* = \rho^* v_j,\ \Phi_A^* = -T_{jA},\ f^* = \rho^* b_j, \end{gathered}$$

and *thus, the theorem of momentum in Lagrangian variables* becomes

$$\frac{\partial \rho^* v_j}{\partial t} - \frac{\partial T_{jA}}{\partial X_A} = \rho^* b_j, \tag{5.48}$$

where we have introduced *the first Piola–Kirchhoff stress tensor*:

$$T_{jA} = t_{ij} F_{iA}^C \quad \Longleftrightarrow \quad \mathbf{T} = \mathbf{t}\mathbf{F}^C, \tag{5.49}$$

so that (5.48) is entirely similar to the corresponding Eulerian momentum balance law (5.15).

Note that $\mathbf{T}$ is not symmetric. In fact,

$$\mathbf{T}^T = \mathbf{F}^{CT}\mathbf{t} \neq \mathbf{T}.$$

Furthermore, (5.49) can be rewritten as

$$\mathbf{T} = J\mathbf{t}\left(\mathbf{F}^{-1}\right)^T = J\mathbf{t}\left(\mathbf{F}^T\right)^{-1} \tag{5.50}$$

from which the inverse formula follows

$$\mathbf{t} = \frac{1}{J}\mathbf{T}\,\mathbf{F}^T \quad \Longleftrightarrow \quad t_{ij} = \frac{1}{J} T_{iA} F_{jA}. \tag{5.51}$$

Since $\mathbf{t} \in \mathcal{Sym}$, we have the condition

$$\mathbf{T}\,\mathbf{F}^T = \mathbf{F}\,\mathbf{T}^T. \tag{5.52}$$

5.4.7 Boundary Conditions in Lagrangian Variables

Taking into account Eq. (2.32), the boundary condition (5.17) becomes

$$\mathbf{t}\mathbf{F}^C\mathbf{n}^* = \mathbf{f}\,J\sqrt{\mathbf{C}^{-1}\mathbf{n}^*\cdot\mathbf{n}^*}. \tag{5.53}$$

Given

$$\mathbf{f}^* = \mathbf{f}J\sqrt{\mathbf{C}^{-1}\mathbf{n}^*\cdot\mathbf{n}^*}, \tag{5.54}$$

Eq. (5.53) becomes

$$\mathbf{T}\mathbf{n}^* = \mathbf{f}^* \quad \Longleftrightarrow \quad T_{iA}n_A^* = f_i^*, \tag{5.55}$$

which is entirely similar to the Eulerian formulation. However, note in Eq. (5.54) how, in addition to the surface force, the deformation that the body undergoes and the normal to the surface $\mathbf{n}^*$ are also involved.

5.4.8 Balance Laws of Energy in Lagrangian Variables

Similarly to what was done for momentum, if we now consider

$$\Psi = \frac{\rho v^2}{2} + \rho e\,; \qquad \Phi_i = -t_{ij}v_j + q_i\,; \qquad f = \rho b_i v_i + r$$
$$\Longrightarrow$$
$$\Psi^* = \tfrac{1}{2}\rho^* v^2 + \rho^* e\,;\ \Phi_A^* = -T_{iA}v_i + Q_A\,;\ f^* = \rho^* b_i v_i + r^*,$$

we obtain from Eq. (5.47) the corresponding *balance law of energy in Lagrangian variables*:

$$\frac{\partial}{\partial t}\left(\frac{1}{2}\rho^* v^2 + \rho^* e\right) + \frac{\partial}{\partial X_A}\left(Q_A - T_{iA}v_i\right) = \rho^* b_i v_i + r^*, \tag{5.56}$$

where we have indicated by $\mathbf{Q}$ the *Piola–Kirchhoff-type heat flux*

$$Q_A = q_i F_{iA}^C \quad \Longleftrightarrow \quad \mathbf{Q} = \mathbf{F}^{CT}\mathbf{q} \tag{5.57}$$

and by

$$r^* = r\frac{\rho^*}{\rho}. \tag{5.58}$$

Regarding the conservation of mass, the Lagrangian formulation (5.47) becomes trivial, as we have

$$\Psi = \rho\,; \qquad \Phi_i = 0\,; \qquad f = 0$$
$$\Longrightarrow$$
$$\Psi^* = \rho^*\,; \quad \Phi_A^* = 0\,; \quad f^* = 0.$$

In fact, in this case, the conservation law of mass in Lagrangian variables is not described by a differential equation, but as we have seen in Eq. (5.3), it translates into a local expression (5.3) $\rho = \rho^*/J$.

In conclusion, the system of thermomechanical laws in Lagrangian variables is

$$\begin{cases} \rho = \rho^*/J \\ \dfrac{\partial \rho^* v_j}{\partial t} - \dfrac{\partial T_{jA}}{\partial X_A} = \rho^* b_j \\ \dfrac{\partial}{\partial t}\left(\rho^* \dfrac{v^2}{2} + \rho^* e\right) + \dfrac{\partial}{\partial X_A}\left(Q_A - T_{iA} v_i\right) = \rho^* b_i v_i + r^*. \end{cases} \tag{5.59}$$

5.5 Physical Interpretation of the First Piola–Kirchhoff Tensor

The physical interpretation of the first Piola–Kirchhoff tensor can be derived from the following considerations. The resultant of contact forces acting on a subvolume ΔV with boundary $\Delta\Sigma$ is given by

$$\int_{\Delta\Sigma} \mathbf{t}_n \, d\Sigma = \int_{\Delta\Sigma} \mathbf{t}\cdot\mathbf{n}\, d\Sigma.$$

However, from Eq. (5.42) $\mathbf{n}\, d\Sigma = \mathbf{F}^C \mathbf{n}^*\, d\Sigma^*$.

Therefore, the integral of the contact forces over the surface $\Delta\Sigma$ can be rewritten as an integral over the surface $\Delta\Sigma^*$, resulting in

$$\int_{\Delta\Sigma} \mathbf{t}_n \, d\Sigma = \int_{\Delta\Sigma^*} \mathbf{t}\mathbf{F}^C \mathbf{n}^* \, d\Sigma^* = \int_{\Delta\Sigma^*} \mathbf{T}\mathbf{n}^* \, d\Sigma^*,$$

where $\mathbf{T}$ represents the Piola–Kirchhoff stress tensor.

By denoting

$$\mathbf{T}_{n^*} = \mathbf{T}\mathbf{n}^*,$$

we can finally write

$$\int_{\Delta\Sigma} \mathbf{t}_n \, d\Sigma = \int_{\Delta\Sigma^*} \mathbf{T}_{n^*} \, d\Sigma^*. \tag{5.60}$$

From Eq. (5.60), it can be concluded that $\mathbf{T}_{n^*}$ should be considered as the specific stress required on $d\Sigma^*$ in order for the resultant on $\Delta\Sigma^*$ to be equal to that of the contact forces on $\Delta\Sigma$.

5.5.1 *Second Piola–Kirchhoff Tensor*

Let us introduce the *second Piola–Kirchhoff tensor* $\mathbf{S}$, which is related to the first tensor by the formula:

$$\mathbf{T} = \mathbf{F}\mathbf{S}. \tag{5.61}$$

The following relationships between the stress tensors can be easily verified (see (5.61), (5.49), and (5.51)):

$$\mathbf{S} = \mathbf{F}^{-1}\mathbf{T} \tag{5.62}$$

$$\mathbf{t} = \frac{1}{J}\,\mathbf{F}\,\mathbf{S}\,\mathbf{F}^T \tag{5.63}$$

$$\mathbf{S} = J\mathbf{F}^{-1}\mathbf{t}\mathbf{F}^{-1T}. \tag{5.64}$$

From (5.64), it immediately follows that $\mathbf{S} = \mathbf{S}^T$, which means that the second Piola–Kirchhoff tensor is symmetric.

5.5.2 Power of Internal Forces in Terms of Piola–Kirchhoff Tensors

The power of internal forces (5.22) can be expressed in terms of the Piola–Kirchhoff tensors, particularly the second tensor.

Recalling Eq. (3.6), we have

$$\nabla \mathbf{v} = \dot{\mathbf{F}}\mathbf{F}^{-1},$$

and thus from (3.3),

$$\mathbf{D} = \frac{1}{2}\left(\dot{\mathbf{F}}\mathbf{F}^{-1} + \mathbf{F}^{-1T}\dot{\mathbf{F}}^{T}\right). \tag{5.65}$$

Using (5.22) and (5.51), we have

$$\mathcal{P}^{(i)} = -\mathbf{t}\cdot\mathbf{D} = -\mathrm{tr}(\mathbf{tD}) = -\frac{1}{J}\mathrm{tr}(\mathbf{T}\mathbf{F}^{T}\mathbf{D}),$$

and from (5.65), we obtain

$$\mathcal{P}^{(i)} = -\frac{1}{2J}\mathrm{tr}(\mathbf{T}\mathbf{F}^{T}\dot{\mathbf{F}}\mathbf{F}^{-1}) - \frac{1}{2J}\mathrm{tr}(\mathbf{T}\mathbf{F}^{T}\mathbf{F}^{-1T}\dot{\mathbf{F}}^{T}).$$

Taking into account (5.52), we have

$$\mathcal{P}^{(i)} = -\frac{1}{2J}\mathrm{tr}(\mathbf{F}\mathbf{T}^{T}\dot{\mathbf{F}}\mathbf{F}^{-1}) - \frac{1}{2J}\mathrm{tr}(\mathbf{T}\dot{\mathbf{F}}^{T}),$$

and using the commutation property $\mathrm{tr}\,(\mathbf{AB}) = \mathrm{tr}\,(\mathbf{BA})$, we obtain

$$\mathcal{P}^{(i)} = -\frac{1}{2J}\,\mathrm{tr}\left(\mathbf{T}^{T}\dot{\mathbf{F}}\right) - \frac{1}{2J}\,\mathrm{tr}\left(\mathbf{T}\dot{\mathbf{F}}^{T}\right).$$

Finally, using the property $\mathrm{tr}\,\mathbf{A} = \mathrm{tr}\,\mathbf{A}^{T}$, we have

$$\mathcal{P}^{(i)} = -\frac{1}{J}\,\mathrm{tr}\left(\mathbf{T}\dot{\mathbf{F}}^{T}\right)$$

or

$$\mathcal{P}^{(i)} = -\frac{1}{J}\mathbf{T}\cdot\dot{\mathbf{F}}.$$

Denoting by $\mathcal{P}_*^{(i)}$ the corresponding power density of internal forces in C^* such that

$$\int_V \mathcal{P}^{(i)} dV = \int_{V^*} \mathcal{P}_*^{(i)} dV^*,$$

we obtain

$$\mathcal{P}_*^{(i)} = J\mathcal{P}^{(i)}, \tag{5.66}$$

and thus we finally have the *power density of internal forces in terms of the first Piola–Kirchhoff tensor*:

$$\mathcal{P}_*^{(i)} = -\mathbf{T} \cdot \dot{\mathbf{F}}. \tag{5.67}$$

By inserting Eq. (5.61), we obtain (taking into account Theorems 1.13.2 and 1.13.1, (1.49), and (1.46))

$$\mathcal{P}_*^{(i)} = -\mathbf{FS} \cdot \dot{\mathbf{F}} = -\operatorname{tr}\left(\mathbf{FS}\dot{\mathbf{F}}^T\right) = -\operatorname{tr}\left(\mathbf{S}\dot{\mathbf{F}}^T\mathbf{F}\right) = -\frac{1}{2}\operatorname{tr}\left(\mathbf{S}\left(\dot{\mathbf{F}}^T\mathbf{F} + \mathbf{F}^T\dot{\mathbf{F}}\right)\right).$$

From Eq. (2.10), we have

$$\dot{\mathbf{C}} = \dot{\mathbf{F}}^T\mathbf{F} + \mathbf{F}^T\dot{\mathbf{F}},$$

and therefore

$$\mathcal{P}_*^{(i)} = -\frac{1}{2}\operatorname{tr}\left(\mathbf{S}\dot{\mathbf{C}}\right) = -\frac{1}{2}\mathbf{S} \cdot \dot{\mathbf{C}}.$$

Finally, from Eq. (2.12), we have

$$\dot{\mathbf{E}} = \frac{1}{2}\dot{\mathbf{C}},$$

and therefore *the power density of internal forces expressed in terms of the second Piola–Kirchhoff tensor becomes simply*

$$\mathcal{P}_*^{(i)} = -\mathbf{S} \cdot \dot{\mathbf{E}}. \tag{5.68}$$

We define the *virtual work of internal forces* in Lagrangian variables per unit mass as

$$\delta l_*^{(i)} = \frac{1}{\rho^*}\mathcal{P}_*^{(i)} dt. \tag{5.69}$$

From Eqs. (5.67) and (5.68), we derive

$$\delta l_{*}^{(i)} = -\frac{1}{\rho^{*}}\mathbf{T} \cdot d\mathbf{F} = -\frac{1}{\rho^{*}}\mathbf{S} \cdot d\mathbf{E}. \tag{5.70}$$

The second equation in (5.70) is very expressive as it clearly shows that *the work of internal forces is due to the variation of deformation of the continuum and is zero for rigid transformations.*

Chapter 6
Constitutive Equations

Abstract The differential system of balance laws, both in Eulerian form (5.31) and in Lagrangian forms (5.48) and (5.56), contains more unknown functions than equations. Therefore, it is necessary to *close the system*, i.e., balance the number of unknowns with the number of equations, by adding new equations that characterize the constitution of the materials. These equations are called for this reason *constitutive equations*.

6.1 General Principles for Constitutive Laws

Constitutive laws, if they have to represent a real physical property of the material, must also satisfy general universal principles:

1. The *material indifference* (or *objectivity*) principle
2. The *entropy* principle

6.1.1 Principle of Material Indifference

The *principle of material indifference* requires that a constitutive law should not depend on the observer.

Referring to Fig. 6.1, consider two observers corresponding to the bases $\{\mathbf{e}_i\}$ and $\{\mathbf{e}'_i\}$, and let $\mathbf{R}$ be the corresponding rotation.

Instead of rotating the observer, we can think that the observer remains the same, and the continuum has changed from the configuration C to configuration C' through a rigid transformation obtained by the inverse rotation $\mathbf{R}^T$.

The transition from C^* to C' corresponds to a Jacobian matrix $\mathbf{F}'$ that differs from $\mathbf{F}$ by the rotation $\mathbf{R}^T$:

$$\mathbf{F}' = \mathbf{R}^T\mathbf{F}. \tag{6.1}$$

© The Author(s), under exclusive license to Springer Nature Switzerland AG 2024

T. Ruggeri, *Introduction to the Thermomechanics of Continua and Hyperbolic Systems*, La Matematica per il 3+2 167,
https://doi.org/10.1007/978-3-031-69951-1_6

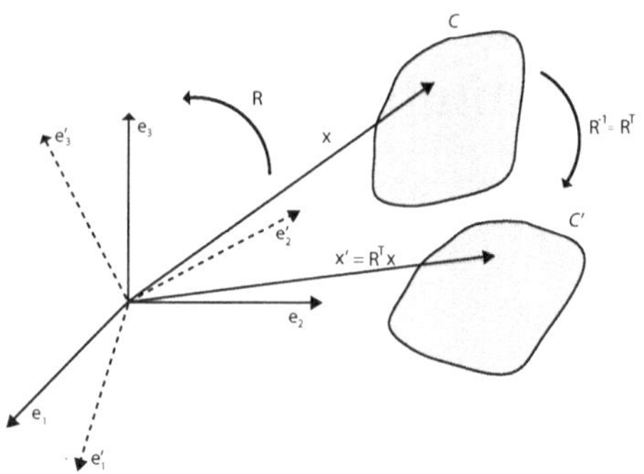

Fig. 6.1 Principle of material indifference

Indeed, considering the generic position vector as $\mathbf{x}'$ in C', we have $\mathbf{x}' = \mathbf{R}^T\mathbf{x}$. Therefore,

$$\mathbf{F}' = \frac{\partial \mathbf{x}'}{\partial \mathbf{X}} = \frac{\partial \mathbf{x}'}{\partial \mathbf{x}}\frac{\partial \mathbf{x}}{\partial \mathbf{X}} = \mathbf{R}^T\mathbf{F}.$$

The constitutive state in C' should be the same as that in C, given that the transition from C to C' effectively corresponds to a change of the observer. Therefore, the principle of material indifference requires, for example, that the stress tensor $\mathbf{t}'$ in C' is equal to that in C up to the similarity transformation:

$$\mathbf{t}' = \mathbf{R}^T\mathbf{t}\mathbf{R}. \tag{6.2}$$

In the case of scalar quantities such as internal energy, the principle of material indifference requires that

$$e' = e. \tag{6.3}$$

In other words, *the principle of material indifference states that if we transition* from *C to C' through* (6.1), *the corresponding constitutive quantities (scalars, vectors, and tensors) should be the same as those in C up to their respective similarity transformations.*

We will see the consequences of the principle of material indifference in the case of elastic and thermoelastic bodies shortly.

6.1.2 The Entropy Principle

The principle of entropy characterizes the modern view of the second law of thermodynamics, reflecting the experience that thermal processes are dissipative and irreversible. For simplicity, let us consider the case of the absence of mass forces and heat sources.

The principle of entropy consists of the following statements based on axioms:

(i) *There exists an additive and objective scalar quantity called entropy.*
(ii) *The entropy density and the entropy flux are constitutive functions to be determined.*
(iii) *The production of entropy is nonnegative for all thermodynamic processes.*

Axiom (i) states that the body possesses an additive quantity, entropy, similar to mass and energy. Just like other additive quantities, there is an entropy balance equation (see (5.29)):

$$\frac{\partial \rho S}{\partial t} + \frac{\partial}{\partial x_i}(\rho S\, v_i + \Phi_i) = \Sigma. \tag{6.4}$$

Here, S is the entropy density, $\mathbf{\Phi} \equiv (\Phi_i)$ is the entropy flux, and Σ is the entropy production density.

Axiom (ii) states that both S and Φ_i are constitutive quantities to be determined, functions of the field variables. In the classical approch (valid only for so-called *simple materials*), the formulation due to Clausius–Duhem is accepted, where the entropy flux is given by the heat flux divided by the absolute temperature ϑ:

$$\mathbf{\Phi} = \frac{\mathbf{q}}{\vartheta}. \tag{6.5}$$

Finally, axiom (iii) requires that the entropy production is nonnegative for any process:

$$\Sigma \geq 0.$$

In the present text, we will write the principle of entropy in the CLAUSIUS–DUHEM form:

$$\frac{\partial \rho S}{\partial t} + \frac{\partial}{\partial x_i}\left(\rho S\, v_i + \frac{q_i}{\vartheta}\right) = \Sigma \geq 0. \tag{6.6}$$

If there is the presence of radiation due to a heat source r, the principle remains essentially the same, but it should be taken into account that the heat source also contributes to the entropy production as r/ϑ. Therefore, the Clausius–Duhem inequality (6.6) becomes

$$\frac{\partial \rho S}{\partial t} + \frac{\partial}{\partial x_i}\left(\rho S\, v_i + \frac{q_i}{\vartheta}\right) = \frac{r}{\vartheta} + \Sigma, \qquad \Sigma \geq 0. \tag{6.7}$$

In other words, in addition to the entropy production due to the heat sources r as the system is not isolated, the second law requires that there is an additional nonnegative entropy production $\Sigma \geq 0$ during every thermodynamic process.

In the case of Lagrangian variables (as mentioned in Sect. 5.4.5), Eq. (6.7) becomes

$$\frac{\partial \rho^* S}{\partial t} + \frac{\partial}{\partial X_A}\left(\frac{Q_A}{\vartheta}\right) = \frac{r^*}{\vartheta} + \Sigma^*, \quad \Sigma^* \geq 0, \tag{6.8}$$

where $r^* = \frac{\rho^*}{\rho} r$ and $\Sigma^* = \frac{\rho^*}{\rho}\Sigma$.

In the upcoming sections, we will discuss the consequences of Eqs. (6.8) and (6.7) in the cases of elasticity and fluids, respectively.

Historical Remark For a long time, constitutive equations were assigned without a precise selection criterion. The inequality of entropy was only seen as an indication of a temporal arrow, denoting the irreversibility of the process. B. Coleman and W. Noll [12] were the first to propose to use the second law of thermodynamics in the form of the Clausius–Duhem inequality as a selection criterion for admissible constitutive equations [12]. Later, I. Müller [13], through considerations of kinetic theory, realized that for non-simple materials (such as mixtures of gases), it was too restrictive to assume the entropy flux of the Clausius–Duhem type (6.5). Müller proposed a principle of entropy in its general form (6.4), where the admissible entropy flux also needs to be determined [13]. The readers interested in the historical aspects of thermodynamics can refer to the book by I. Müller [14].

Chapter 7
Elasticity and Thermoelasticity

Abstract In this chapter, we will describe *elastic* and *thermoelastic* bodies and study the consequences of the *principles of material indifference* and *entropy*.

7.1 Elastic Bodies

We will begin by considering the purely mechanical case where the temperature is constant, and the balance laws reduce to only the balance law of momentum.

A body is said to be elastic when the stress tensor **t** *depends solely on the current value of* **F**:

$$\mathbf{t} \equiv \mathbf{t}(\mathbf{F}) . \tag{7.1}$$

Recalling the relationship between **F** and the displacement gradient (2.7), we have that the system (5.48), with the constitutive equation (7.1), forms a second-order differential system for the components of the displacement vector **u** as functions of **X** and t:

$$\frac{\partial^2 \rho^* u_i}{\partial t^2} - \frac{\partial T_{iA}\left(\frac{\partial u_k}{\partial X_B}\right)}{\partial X_A} = \rho^* b_i . \tag{7.2}$$

Instead of writing the second-order system (7.2), it may be convenient to consider the equivalent first-order system:

$$\begin{cases} \rho^* \dfrac{\partial v_i}{\partial t} - \dfrac{\partial T_{iA}(F_{kB})}{\partial X_A} = \rho^* b_i \\[2ex] \dfrac{\partial F_{iA}}{\partial t} - \dfrac{\partial v_i}{\partial X_A} = 0 \end{cases} \tag{7.3}$$

© The Author(s), under exclusive license to Springer Nature Switzerland AG 2024

T. Ruggeri, *Introduction to the Thermomechanics of Continua and Hyperbolic Systems*, La Matematica per il 3+2 167,
https://doi.org/10.1007/978-3-031-69951-1_7

in the unknowns $\mathbf{v} \equiv \mathbf{v}(\mathbf{X},t)$ and $\mathbf{F} \equiv \mathbf{F}(\mathbf{X},t)$. The second equation in (7.3) arises from the compatibility condition between

$$F_{iA} = \frac{\partial x_i}{\partial X_A} \quad \text{and} \quad v_i = \frac{\partial x_i}{\partial t}.$$

7.1.1 Consequences of the Principle of Material Indifference in the Elastic Case

We want to prove that for an elastic body $\mathbf{t} \equiv \mathbf{t}(\mathbf{F})$, the following result holds:

Theorem 7.1.1 *A necessary and sufficient condition for the Principle of Material Indifference to be satisfied for an elastic material is that the second Piola–Kirchhoff tensor depends on* $\mathbf{F}$ *only through the deformation matrix* $\mathbf{E}$*:*

$$\mathbf{S} \equiv \mathbf{S}(\mathbf{E}) \quad \textit{or equivalently} \quad \mathbf{S} \equiv \mathbf{S}(\mathbf{C}).$$

Proof Under the assumption of elasticity, the condition (5.65) expressing the Principle of Material Indifference becomes

$$\mathbf{t}(\mathbf{F}') = \mathbf{R}^T \mathbf{t}(\mathbf{F})\mathbf{R} \quad \forall\, \mathbf{R} \in \mathcal{Rot} \qquad \text{where } \mathbf{F}' = \mathbf{R}^T \mathbf{F},$$

which implies

$$\mathbf{t}\left(\mathbf{R}^T \mathbf{F}\right) = \mathbf{R}^T \mathbf{t}(\mathbf{F})\,\mathbf{R} \qquad \forall\, \mathbf{R} \in \mathcal{Rot}. \tag{7.4}$$

By substituting (7.4) into (5.63), we obtain

$$\frac{1}{J}\mathbf{R}^T \mathbf{F}\mathbf{S}\left(\mathbf{R}^T \mathbf{F}\right)\mathbf{F}^T \mathbf{R} = \frac{1}{J}\mathbf{R}^T \mathbf{F}\mathbf{S}(\mathbf{F})\,\mathbf{F}^T \mathbf{R} \qquad \forall \mathbf{R} \in \mathcal{Rot},$$

which can be written as

$$\mathbf{S}\left(\mathbf{R}^T \mathbf{F}\right) = \mathbf{S}(\mathbf{F}) \qquad \forall \mathbf{R} \in \mathcal{Rot}. \tag{7.5}$$

Using the polar decomposition theorem

$$\mathbf{F} = \hat{\mathbf{R}}\mathbf{U} \tag{7.6}$$

and requiring (7.5) to hold even for $\mathbf{R} = \hat{\mathbf{R}}$, we immediately obtain

$$\mathbf{S}(\mathbf{U}) = \mathbf{S}(\mathbf{F}),$$

which implies that $\mathbf{S}$ depends on $\mathbf{F}$ only through the deformation tensor $\mathbf{U}$, or equivalently, that $\mathbf{S}$ depends on $\mathbf{C}$ ($\mathbf{C} = \mathbf{U}^2$) or $\mathbf{E}$ ($\mathbf{E} = \frac{1}{2}(\mathbf{C} - \mathbf{I})$):

$$\mathbf{S} \equiv \mathbf{S}\,(\mathbf{E})\,.$$

□

From Eq. (5.63), we have that the dependence of $\mathbf{t}$ on $\mathbf{F}$ must be given by

$$\mathbf{t}(\mathbf{F}) = \frac{1}{J}\,\mathbf{F}\,\mathbf{S}(\mathbf{E})\,\mathbf{F}^T, \tag{7.7}$$

while the first Piola–Kirchhoff stress tensor (5.61) is given by

$$\mathbf{T}(\mathbf{F}) = \mathbf{F}\mathbf{S}(\mathbf{E}).$$

7.2 Thermoelastic Bodies

A body is said to be thermoelastic when the constitutive quantities depend not only on $\mathbf{F}$ *but also on the temperature* ϑ *and possibly on its gradient.* In the usual case, we have

$$\begin{cases} \mathbf{t} \equiv \mathbf{t}\,(\mathbf{F},\,\vartheta) \\ e \equiv e\,(\mathbf{F},\,\vartheta) \\ \mathbf{q} = -\chi\,(\mathbf{F},\vartheta)\,\nabla\vartheta. \end{cases} \tag{7.8}$$

The third equation in (7.8) represents the so called *Fourier's constitutive equation*, which assumes that the heat flux is proportional to the temperature gradient. The function χ is called thermal conductivity.

7.2.1 Principle of Material Indifference in Thermoelasticity

Theorem 7.2.1 *A necessary and sufficient condition for the Principle of Material Indifference to be satisfied for a thermoelastic material is that the second Piola–Kirchhoff tensor, the internal energy, and the thermal conductivity depend on* $\mathbf{F}$ *only through the deformation matrix* $\mathbf{E}$*:*

$$\mathbf{S} \equiv \mathbf{S}\,(\mathbf{E},\vartheta)\,, \quad e \equiv e(\mathbf{E},\,\vartheta), \quad \chi \equiv \chi(\mathbf{E},\,\vartheta). \tag{7.9}$$

Proof The proof regarding $\mathbf{S}$ remains the same as in the elastic case, and we have

$$\mathbf{S}\,(\mathbf{F},\,\vartheta) = \mathbf{S}\,(\mathbf{E},\,\vartheta)\,. \tag{7.10}$$

As for the scalars $e(\mathbf{F}, \vartheta)$ and $\chi(\mathbf{F}, \vartheta)$, it is immediate that they must depend on $\mathbf{F}$ only through the deformation $\mathbf{E}$. In fact, we require (6.3), which states

$$e(\mathbf{F}', \vartheta) = e(\mathbf{F}, \vartheta)$$

implying

$$e(\mathbf{R}^T\mathbf{F}, \vartheta) = e(\mathbf{F}, \vartheta) \quad \forall \mathbf{R} \in \mathcal{R}ot. \tag{7.11}$$

By choosing $\mathbf{R}$ to be coincident with the rotation $\hat{\mathbf{R}}$ appearing in (7.6), we have from (7.11)

$$e(\mathbf{U}, \vartheta) = e(\mathbf{F}, \vartheta). \tag{7.12}$$

Therefore, (7.12) shows that the functional dependence of the scalars on $\mathbf{F}$ is only through the deformation $\mathbf{U}$, which is equivalent to $\mathbf{C}$ or $\mathbf{E}$, and the theorem is proven. □

In conclusion, the Principle of Material Indifference requires that the constitutive equations (7.8) satisfy

$$\begin{cases} \mathbf{t}\,(\mathbf{F},\vartheta) = \frac{1}{J}\,\mathbf{F}\,\mathbf{S}\,(\mathbf{E},\,\vartheta)\;\mathbf{F}^T \\ e\,(\mathbf{F},\vartheta) = e(\mathbf{E},\vartheta) \\ \mathbf{q}\,(\mathbf{F},\,\vartheta,\,\nabla\vartheta) = -\chi\,(\mathbf{E},\vartheta)\,\nabla\vartheta. \end{cases} \tag{7.13}$$

7.2.2 *Field Equations of Thermoelasticity*

Let us rewrite the differential system of thermoelasticity as a first-order system:

$$\begin{cases} \rho^* \dfrac{\partial v_i}{\partial t} - \dfrac{\partial T_{iA}}{\partial X_A} = \rho^* b_i \\ \dfrac{\partial F_{iA}}{\partial t} - \dfrac{\partial v_i}{\partial X_A} = 0 \\ \rho^* \dfrac{\partial}{\partial t}\left(\dfrac{v^2}{2} + e\right) - \dfrac{\partial}{\partial X_A}\left(T_{iA} v_i - Q_A\right) = \rho^* b_i v_i + r^*. \end{cases} \tag{7.14}$$

Multiplying Eq. (7.14)$_1$ scalarly by $\mathbf{v}$, we have

$$\rho^* v_i \frac{\partial v_i}{\partial t} - v_i \frac{\partial T_{iA}}{\partial X_A} = \rho^* b_i v_i \quad \Rightarrow \quad \rho^* \frac{\partial}{\partial t}\left(\frac{v^2}{2}\right) - v_i \frac{\partial T_{iA}}{\partial X_A} = \rho^* b_i v_i. \tag{7.15}$$

Subtracting equation (7.15) from (7.14)$_3$, we obtain the *evolution law (not balance law) for internal energy*:

$$\rho^* \frac{\partial e}{\partial t} - T_{iA} \frac{\partial v_i}{\partial X_A} + \frac{\partial Q_A}{\partial X_A} = r^*. \tag{7.16}$$

For *classical solutions*, Eq. (7.16) is equivalent to (7.14)$_3$.[1] Taking into account (7.8), Eq. (7.16) becomes

$$\rho^* \left(\frac{\partial e}{\partial F_{iA}} \frac{\partial F_{iA}}{\partial t} + \frac{\partial e}{\partial \vartheta} \frac{\partial \vartheta}{\partial t}\right) - T_{iA} \frac{\partial v_i}{\partial X_A} + \frac{\partial Q_A}{\partial X_A} = r^*. \tag{7.17}$$

By substituting Eq. (7.14)$_2$ into (7.17), we obtain the *evolution law for temperature*:

$$\frac{\partial \vartheta}{\partial t} = \frac{1}{\frac{\partial e}{\partial \vartheta}} \left\{\left(\frac{T_{iA}}{\rho^*} - \frac{\partial e}{\partial F_{iA}}\right) \frac{\partial v_i}{\partial X_A} - \frac{1}{\rho^*} \frac{\partial Q_A}{\partial X_A} + \frac{r^*}{\rho^*}\right\}. \tag{7.18}$$

Under differentiability conditions, every solution of (7.14) is also a solution of the system formed by the field equations:

$$\begin{cases} \rho^* \dfrac{\partial v_i}{\partial t} = \dfrac{\partial T_{iA}}{\partial X_A} + \rho^* b_i \\[2ex] \dfrac{\partial F_{lA}}{\partial t} = \dfrac{\partial v_l}{\partial X_A} \\[2ex] \dfrac{\partial \vartheta}{\partial t} = \dfrac{1}{\dfrac{\partial e}{\partial \vartheta}} \left\{\left(\dfrac{T_{iA}}{\rho^*} - \dfrac{\partial e}{\partial F_{iA}}\right) \dfrac{\partial v_i}{\partial X_A} - \dfrac{1}{\rho^*} \dfrac{\partial Q_A}{\partial X_A} + \dfrac{r^*}{\rho^*}\right\}. \end{cases} \tag{7.19}$$

Note that by using (7.13)$_3$ and (5.57), we have

$$Q_A = -\chi F^C_{iA} \frac{\partial \vartheta}{\partial x_i} = -\chi F^C_{iA} \frac{\partial \vartheta}{\partial X_B} \frac{\partial X_B}{\partial x_i} = -\chi J F^{-1}_{Ai} F^{-1}_{Bi} \frac{\partial \vartheta}{\partial X_B}.$$

Thus,

$$\mathbf{Q} = -\chi J \mathbf{F}^{-1} \left(\mathbf{F}^{-1}\right)^T \operatorname{Grad} \vartheta,$$

[1] The concept of classical and weak solutions will be clarified later in Chap. 11.

where

$$\text{Grad}\,\vartheta \equiv \left(\frac{\partial\vartheta}{\partial X_1}, \frac{\partial\vartheta}{\partial X_2}, \frac{\partial\vartheta}{\partial X_3}\right).$$

Finally, using (2.10), we have

$$\mathbf{Q} = -\chi(\mathbf{E}, \vartheta)\; J\mathbf{C}^{-1}\text{Grad}\,\vartheta, \tag{7.20}$$

which represents *Fourier's law in Lagrangian variables.*

Therefore, given the constitutive functions (7.9), we have the heat flux given by (7.20) and the PIOLA–KIRCHHOFF tensor given by (5.53), expressed in terms of $\mathbf{F}$ and ϑ. Thus, the system has the same number of field equations (7.19) as the number of the unknowns:

$$\mathbf{v} \equiv \mathbf{v}(\mathbf{X}, t), \;\; \mathbf{F} \equiv \mathbf{F}(\mathbf{X}, t), \;\; \text{and}\; \vartheta \equiv \vartheta(\mathbf{X}, t).$$

In the next section, we will see that further strong restrictions on the constitutive equations come from the principle of entropy.

7.2.3 *Consequences of the Entropy Principle in Thermoelasticity*

We consider the consequences of the entropy principle in thermoelasticity. We introduce the entropy density, denoted by $S(\mathbf{F}, \vartheta)$, which is another constitutive function to be determined

$$S \equiv S\,(\mathbf{F}, \vartheta)\,. \tag{7.21}$$

The principle of material indifference implies that the functional dependence on $\mathbf{F}$ is through the deformation $\mathbf{E}$; thus $S(\mathbf{F}, \vartheta) = S(\mathbf{E}, \vartheta)$.

The following theorem states the necessary and sufficient conditions for the entropy principle to hold in thermoelastic materials:

Theorem 7.2.2 *For the entropy principle to be satisfied in thermoelastic materials, it is necessary and sufficient that there exists a free energy function,* $\psi(\mathbf{E}, \vartheta)$, *dependent on the deformation and temperature, such that*

$$S = -\frac{\partial\psi}{\partial\vartheta}, \quad \mathbf{S} = \rho^*\frac{\partial\psi}{\partial\mathbf{E}}, \quad e = \psi - \vartheta\frac{\partial\psi}{\partial\vartheta}. \tag{7.22}$$

Additionally, the thermal conductivity function $\chi(\mathbf{E}, \vartheta)$ *must be nonnegative:* $\chi(\mathbf{E}, \vartheta) \geq 0$.

Proof By considering Eqs. (6.8) and (7.21), we have

$$\rho^*\left(\frac{\partial S}{\partial F_{iA}}\frac{\partial F_{iA}}{\partial t}+\frac{\partial S}{\partial \vartheta}\frac{\partial \vartheta}{\partial t}\right)+\frac{1}{\vartheta}\frac{\partial Q_A}{\partial X_A}-\frac{1}{\vartheta^2}Q_A\frac{\partial \vartheta}{\partial X_A}-\frac{r^*}{\vartheta}=\Sigma^*\geq 0.$$

Substituting the temporal derivatives $\frac{\partial F_{iA}}{\partial t}$ and $\frac{\partial \vartheta}{\partial t}$ from Eqs. $(7.19)_2$ and $(7.19)_3$ and considering (7.20), we obtain

$$\rho^*\left\{\frac{\partial S}{\partial F_{iA}}+\frac{\frac{\partial S}{\partial \vartheta}}{\frac{\partial e}{\partial \vartheta}}\left(\frac{T_{iA}}{\rho^*}-\frac{\partial e}{\partial F_{iA}}\right)\right\}\frac{\partial v_i}{\partial X_A}+\left\{\frac{1}{\vartheta}-\frac{\frac{\partial S}{\partial \vartheta}}{\frac{\partial e}{\partial \vartheta}}\right\}\left(\frac{\partial Q_A}{\partial X_A}-r^*\right)+$$

$$+\frac{1}{\vartheta^2}\chi J\mathbf{C}^{-1}\mathrm{Grad}\,\vartheta\cdot\mathrm{Grad}\,\vartheta=\Sigma^*\geq 0. \tag{7.23}$$

The inequality above must hold for all possible processes. Therefore, for any initial data, along with their spatial derivatives at a generic time t_0, the inequality (7.23) must be satisfied. Considering that the spatial derivatives $\frac{\partial v_i}{\partial X_A}$ and $\frac{\partial Q_A}{\partial X_A}$ in the above equation are linear, while the last term is quadratic, the only way for the inequality to be nonnegative is for the terms in curly braces to be zero and for χ to be a nonnegative function (noting that $J > 0$ and that $\mathbf{C}$, and consequently $\mathbf{C}^{-1}$, are positive definite matrices):

$$\begin{cases}\vartheta\dfrac{\partial S}{\partial \vartheta}=\dfrac{\partial e}{\partial \vartheta},\\[2ex] \vartheta\dfrac{\partial S}{\partial F_{iA}}=\dfrac{\partial e}{\partial F_{iA}}-\dfrac{T_{iA}}{\rho^*},\\[2ex] \chi\geq 0.\end{cases} \tag{7.24}$$

The first two equations in (7.24) are equivalent to the differential form:

$$\vartheta dS=de-\frac{1}{\rho^*}\mathbf{T}\cdot d\mathbf{F}, \tag{7.25}$$

which represents the well-known *Gibbs equation*.

Taking into account equations (5.70) and (7.25) can also be written as

$$\vartheta dS=de+\delta l_*^{(i)}. \tag{7.26}$$

The formula of *Gibbs* (7.26) states that the variation in internal energy is transformed partly into heat ϑdS and partly into the work of internal forces.

It should be noted that in this presentation, the formula of *Gibbs* is a consequence of the entropy principle. In traditional books, Eq. (7.26) is postulated in nonequilibrium and is known as the hypothesis of *local equilibrium*.[2]

Using Eqs. (5.70) and (7.25) can also be written as

$$\vartheta dS = de - \frac{1}{\rho^*}\mathbf{S} \cdot d\mathbf{E}. \tag{7.27}$$

Introducing the free energy function

$$\psi = e - \vartheta S \tag{7.28}$$

and differentiating equation (7.28) while taking Eq. (7.27) into account, we obtain

$$d\psi = -S\, d\vartheta + \frac{1}{\rho^*}\mathbf{S} \cdot d\mathbf{E}. \tag{7.29}$$

Considering $\psi \equiv \psi(\mathbf{E}, \vartheta)$ from Eq. (7.29), we have

$$S = -\frac{\partial \psi}{\partial \vartheta}, \qquad \mathbf{S} = \rho^* \frac{\partial \psi}{\partial \mathbf{E}}, \tag{7.30}$$

and from Eq. (7.28),

$$e = \psi - \vartheta \frac{\partial \psi}{\partial \vartheta}.$$

This completes the proof of the theorem. □

Note from (7.23) that the entropy production is given by

$$\Sigma^* = \frac{\chi J}{\vartheta^2} \mathbf{C}^{-1} \text{Grad}\, \vartheta \cdot \text{Grad}\, \vartheta \geq 0. \tag{7.31}$$

Therefore, if the thermal conductivity is zero, the transformation becomes reversible $\Sigma^* \equiv 0$.

Thus, a thermoelastic body is characterized by only two constitutive functions: the free energy $\psi(\mathbf{E}, \vartheta)$ *and the thermal conductivity* $\chi(\mathbf{E}, \vartheta) \geq 0$.

From (5.61) and $(7.30)_2$, we have that the first Piola–Kirchhoff tensor compatible with the entropy principle is

$$\mathbf{T} = \rho^* \mathbf{F} \frac{\partial \psi}{\partial \mathbf{E}}. \tag{7.32}$$

---

[2] In the case of thermostatic, the formula of *Gibbs* can be derived using a theorem due to CARATHÉODORY.

7.2.4 Isotropic Materials

If the body is isotropic, the function ψ depends on $\mathbf{E}$ through its invariants, so that the property remains the same regardless of the observer's frame. Therefore,

$$\psi \equiv \psi\left(I_1, I_2, I_3, \vartheta\right) \tag{7.33}$$

with I_1, I_2, and I_3 being the principal invariants of $\mathbf{E}$:

$$I_1 = \operatorname{tr}\mathbf{E}, \quad I_2 = \operatorname{tr}\mathbf{E}^C, \quad I_3 = \det\mathbf{E}.$$

From $(7.22)_2$, considering (1.80), we have

$$\mathbf{S} = g_0\,\mathbf{I} + g_1\,\mathbf{E} + g_{-1}\,\mathbf{E}^C, \tag{7.34}$$

where

$$\begin{aligned} g_0 &= \rho^*\left(\psi_{I_1} + I_1\psi_{I_2}\right), \\ g_1 &= -\rho^*\psi_{I_2}, \\ g_{-1} &= \rho^*\psi_{I_3}, \end{aligned}$$

and

$$\psi_{I_1} = \frac{\partial\psi}{\partial I_1}, \quad \psi_{I_2} = \frac{\partial\psi}{\partial I_2}, \quad \psi_{I_3} = \frac{\partial\psi}{\partial I_3}.$$

From the HAMILTON–CAYLEY theorem (1.78),

$$\mathbf{E}^C = \mathbf{E}^2 - I_1\mathbf{E} + I_2\mathbf{I},$$

substituting in (7.34), we *ultimately obtain the stress–strain relationship for a nonlinear isotropic material*:

$$\mathbf{S} = f_0\,\mathbf{I} + f_1\,\mathbf{E} + f_2\,\mathbf{E}^2 \tag{7.35}$$

with *response functions*:

$$\begin{aligned} f_0 &= \rho^*\left(\psi_{I_1} + I_1\psi_{I_2} + I_2\psi_{I_3}\right), \\ f_1 &= -\rho^*\left(\psi_{I_2} + I_1\psi_{I_3}\right), \\ f_2 &= \rho^*\psi_{I_3}. \end{aligned} \tag{7.36}$$

Sometimes, instead of the principal invariants of $\mathbf{E}$, it is convenient to use the invariants of the trace of the powers of $\mathbf{E}$ (1.72) and consider ψ as a function of these:

$$\psi \equiv \psi\left(J_1, J_2, J_3, \vartheta\right), \qquad J_1 = \operatorname{tr}\mathbf{E},\ J_2 = \operatorname{tr}\mathbf{E}^2,\ J_3 = \operatorname{tr}\mathbf{E}^3. \tag{7.37}$$

In this case, considering the derivation formulas (1.73), we have (7.35) with response functions:

$$f_0 = \rho^* \psi_{J_1}, \quad f_1 = 2\rho^* \psi_{J_2}, \quad f_2 = 3\rho^* \psi_{J_3}. \tag{7.38}$$

7.3 Principle of Dissipation in Elasticity

We have seen the consequences of the entropy principle in the case of thermoelasticity. One might wonder what corresponds to the entropy principle in the purely mechanical case.

In the purely mechanical domain, the entropy principle is referred to as the principle of energy dissipation, which essentially states that in a mechanical process, energy either remains conserved or decreases.

In other words, we need to require that every solution of

$$\begin{cases} \rho^* \dfrac{\partial v_i}{\partial t} - \dfrac{\partial T_{iA}}{\partial X_A} = \rho^* b_i \\[2ex] \dfrac{\partial F_{iA}}{\partial t} = \dfrac{\partial v_i}{\partial X_A} \end{cases} \tag{7.39}$$

is also a solution of the energy balance equation with nonpositive energy production:

$$\rho^* \frac{\partial}{\partial t}\left(\frac{v^2}{2} + e\right) - \frac{\partial}{\partial X_A}\left(T_{iA} v_i\right) - \rho^* b_i v_i = \mathcal{E} \leq 0. \tag{7.40}$$

The elastic case corresponds to the ideal case of reversible transformation. In fact, we will prove that $\mathcal{E} = 0$.

Proceeding as in the thermoelastic case, we multiply $(7.39)_1$ by v_i, subtract it from (7.40), and take into account $(7.39)_2$, resulting in

$$\left(\rho^* \frac{\partial e}{\partial F_{iA}} - T_{iA}\right) \frac{\partial v_i}{\partial X_A} = \mathcal{E} \leq 0. \tag{7.41}$$

Considering that, according to the hypothesis of elasticity, T_{iA} depends only on F_{jB} and that the left-hand side is linear in arbitrary spatial derivatives, we necessarily have

$$T_{iA} = \rho^* \frac{\partial e}{\partial F_{iA}}, \tag{7.42}$$

and $\mathcal{E} = 0$. Equation (7.42) is equivalent to

$$de = \frac{1}{\rho^*} \mathbf{T} \cdot d\mathbf{F}$$

Taking into account (5.70),

$$de = -\delta l_*^{(i)},$$

which corresponds to Gibbs' formula (7.26) with $S = constant$. The physical interpretation is that *in the purely mechanical case, the variation of internal energy is transformed into work of internal forces*. From (5.70), we also have

$$de = \frac{1}{\rho^*} \mathbf{S} \cdot d\mathbf{E};$$

hence

$$\mathbf{S} = \rho^* \frac{\partial e}{\partial \mathbf{E}}. \tag{7.43}$$

Thus, the principle of dissipation, which we will continue to refer to as the entropy principle, implies that the stress $\mathbf{S}$ and the internal energy must satisfy (7.43).

In the literature, it is common to introduce the *strain energy density*, denoted by W:

$$W = \rho^* e, \tag{7.44}$$

such that Eqs. (7.42) and (7.43) become

$$\mathbf{T} = \frac{\partial W}{\partial \mathbf{F}}, \qquad \mathbf{S} = \frac{\partial W}{\partial \mathbf{E}}. \tag{7.45}$$

Hence we have the following theorem:

Theorem 7.3.1 *The dissipation principle in elasticity implies the existence of a strain energy density such that stress and deformation are related through (7.45).*

Note that (7.42) can be seen as a limiting case of the results in thermoelasticity (7.24) by setting S as constant and $\chi = 0$.

In the case of isotropic elasticity, similar conclusions to (7.35) hold

$$\mathbf{S} = f_0\,\mathbf{I} + f_1\,\mathbf{E} + f_2\,\mathbf{E}^2 \tag{7.46}$$

with

$$f_0 = W_{I_1} + I_1\,W_{I_2} + I_2\,W_{I_3}, \quad f_1 = -\left(W_{I_2} + I_1\,W_{I_3}\right), \quad f_2 = W_{I_3}, \tag{7.47}$$

if we consider $W \equiv W(I_1, I_2, I_3)$, or

$$f_0 = W_{J_1}, \quad f_1 = 2W_{J_2}, \quad f_2 = 3W_{J_3}, \tag{7.48}$$

if we assign $W \equiv W(J_1, J_2, J_3)$.

If we assume that in the absence of deformation there is also no stress according to (7.46), we must impose

$$f_0(0, 0, 0) = 0. \tag{7.49}$$

Materials for which stress and deformation are related through a potential (7.45) *in the elastic case and* $(7.30)_2$ *in the thermoelastic case are called hyperelastic materials.*

Remark 7 (Historical Remark) For a long time, it was believed that there are more general elastic materials than hyperelastic ones. For example, isotropic materials for which the three response functions appearing in (7.46) are arbitrary functions of the three principal invariants of $\mathbf{E}$. However, as seen in this section and previously in the case of thermoelasticity, the only materials compatible with the entropy principle are hyperelastic materials, and their response functions depend on a single potential (see (7.47) or (7.48) in the elastic case and (7.36) or (7.38) in the thermoelastic case). To my knowledge, the author who first arrived at this conclusion in the elastic case was R. Rivlin [16].

7.3.1 Compressible Mooney–Rivlin Potential

An example of strain energy density used in engineering is the compressible MOONEY–RIVLIN potential [15, 16]. This is a constitutive model used in continuum mechanics to describe the mechanical behavior of elastomeric materials, particularly rubber-like substances. The general form of the strain energy density function can be expressed as

$$W = C_1(\bar{I}_1 - 3) + C_2(\bar{I}_2 - 3) + \frac{1}{D}(J - 1)^2, \tag{7.50}$$

where:

- C_1 and C_2 are material constants associated with the deviatoric response.
- $\bar{I}_1$ and $\bar{I}_2$ are the first and second invariants of the unimodular deformation tensor $\bar{\mathbf{B}}$ defined in (2.37).
- D is a material parameter related to the compressibility of the material.

In this model, the terms involving $\bar{I}_1$ and $\bar{I}_2$ capture the nonlinear elastic response typical of rubber-like materials, while the term involving J accounts for volumetric changes, allowing the model to handle compressible behaviors. If $C_2 = 0$, we obtain a special case of a Mooney–Rivlin model that is called *neo-Hookean* solid.

7.3.2 *Incompressible Hyperelastic Solids*

In many practical problems, the elastic material can be considered incompressible, i.e., $J = 1$. As a consequence of the constraint, there exists a *constraint reaction* that mathematically corresponds to substituting the strain energy density $W(\mathbf{F})$ with

$$W(\mathbf{F}) \quad \Rightarrow \quad W(\mathbf{F}) - p(J - 1), \tag{7.51}$$

where p is an indeterminate Lagrange multiplier that has the physical meaning of *hydrostatic pressure*. In this case, the first and second Piola–Kirchhoff stress tensors, taking into account (1.80) and (5.62), become

$$\mathbf{T} = \frac{\partial W}{\partial \mathbf{F}} - p\,\mathbf{F}^C, \qquad \mathbf{S} = \frac{\partial W}{\partial \mathbf{E}} - p\,\mathbf{C}^{-1}. \tag{7.52}$$

The Mooney–Rivlin strain energy density (7.50) reduces to

$$W = C_1(\bar{I}_1 - 3) + C_2(\bar{I}_2 - 3). \tag{7.53}$$

Remark 8 The Mooney–Rivlin model, although widely used, does not always provide results in agreement with experimental data, especially for very large deformations. The search for a strain energy density function, in both incompressible and compressible cases, remains an active field of research. Many proposals have been made in the literature, among which a very popular model for the incompressible case is the one proposed by Ogden [17].

7.3.3 Nonlinear One-Dimensional Elasticity

In the case of a one-dimensional problem where $\mathbf{u} \equiv (u, 0, 0)$ depends only on a spatial variable $X = X_1$, considering that (2.7) and $(2.15)_4$ become

$$\mathbf{F} \equiv \begin{pmatrix} 1+u_X & 0 & 0 \\ 0 & 1 & 0 \\ 0 & 0 & 1 \end{pmatrix}, \qquad \mathbf{E} \equiv \begin{pmatrix} u_X + \frac{1}{2}{u_X}^2 & 0 & 0 \\ 0 & 0 & 0 \\ 0 & 0 & 0 \end{pmatrix} \tag{7.54}$$

$$J = 1 + u_X, \qquad u_X = \frac{\partial u}{\partial X},$$

from (7.42) and (7.2), we obtain that the field system reduces to a *nonlinear wave equation*:

$$\frac{\partial^2 u}{\partial t^2} - e''(u_X)\frac{\partial^2 u}{\partial X^2} = 0, \qquad \left(e''(u_X) = \frac{\partial^2 e}{\partial u_X^2} \right). \tag{7.55}$$

7.4 Linear Elasticity

Let us assume now that the deformation is small, meaning that the deformation gradient and therefore the strain tensor $\mathbf{E}$ are small enough to neglect the square of their components. To indicate this, we write

$$\text{Grad}\,\mathbf{u} = O(\varepsilon), \qquad \mathbf{E} = O(\varepsilon). \tag{7.56}$$

In this case, the approximate formulas discussed in Sect. 2.12 hold.

Note that from (2.7) and (2.44), we also have

$$\mathbf{F}^{-1} \simeq \mathbf{I} - \text{Grad}\,\mathbf{u}, \qquad \mathbf{F}^C = J\mathbf{F}^{-1T} \simeq (1 + \text{div}\mathbf{u})\mathbf{I} - (\text{Grad}\,\mathbf{u})^{\mathrm{T}}. \tag{7.57}$$

We also assume that the Cauchy stress tensor and the velocity are small as well:

$$\mathbf{t} = O(\varepsilon), \qquad \mathbf{v} = O(\varepsilon).$$

Under these assumptions of linear elasticity, it can be seen immediately from (5.49) and (5.61), taking into account (7.57), that

$$\mathbf{t} \simeq \mathbf{T} \simeq \mathbf{S}. \tag{7.58}$$

Therefore, we have the result that in *a theory of linear elasticity, all stress tensors coincide*.

Furthermore, if the gradients are small and the velocity is also small, the material derivative and the partial derivative with respect to time coincide (see (3.2)) and the spatial derivatives with respect to $\mathbf{X}$ and $\mathbf{x}$ along with the densities:

$$\frac{\partial}{\partial X_A} = F_{iA}\frac{\partial}{\partial x_i} \simeq \frac{\partial}{\partial x_A}, \qquad \frac{d}{dt} = \frac{\partial}{\partial t} + v_i\frac{\partial}{\partial x_i} \simeq \frac{\partial}{\partial t}, \qquad \rho^* = \rho J \simeq \rho.$$

Therefore, the Lagrangian momentum balance equation (5.48) *and the Eulerian balance equation* (5.15) *are equivalent in linear elasticity, and we have*

$$\rho^* \frac{\partial v_i}{\partial t} - \frac{\partial t_{ij}}{\partial x_j} = \rho^* b_i. \tag{7.59}$$

Similar considerations can be made for the energy balance equation.

7.4.1 Linear Isotropic Elasticity Equations

In the isotropic case, considering $W \equiv W(J_1, J_2, J_3)$, we have

$$J_1 = \operatorname{tr}\mathbf{E} = O(\varepsilon), \quad J_2 = \operatorname{tr}\mathbf{E}^2 = O(\varepsilon^2), \quad J_3 = \operatorname{tr}\mathbf{E}^3 = O(\varepsilon^3),$$

and in order to have a linear stress–strain relationship (7.46), we need to take a Taylor expansion of $W \equiv W(J_1, J_2, J_3)$ up to second order. The only possible potential is therefore (the linear term is absent due to the condition (7.49))

$$W = \frac{\nu}{2}J_1^2 + \mu J_2, \tag{7.60}$$

where the constants ν and μ are known as the LAMÉ constants.

Substituting (7.60) into (7.46) and taking into account (7.48) and (7.58), we obtain the stress–strain relationship for linear isotropic elasticity:

$$\mathbf{t} = \nu J_1 \mathbf{I} + 2\mu\mathbf{E},$$

which, using (2.43), can be rewritten in terms of the displacement vector:

$$\mathbf{t} = \nu\nabla\cdot\mathbf{u}\,\mathbf{I} + \mu\left(\nabla\mathbf{u} + (\nabla\mathbf{u})^T\right)$$

or in component form:

$$t_{ij} = \nu \frac{\partial u_k}{\partial x_k} \delta_{ij} + \mu \left(\frac{\partial u_i}{\partial x_j} + \frac{\partial u_j}{\partial x_i} \right).$$

Inserting this into (7.59), we obtain the evolution equations for the displacement vector:

$$\rho \frac{\partial^2 \mathbf{u}}{\partial t^2} - (\nu + \mu) \nabla (\nabla \cdot \mathbf{u}) - \mu \Delta \mathbf{u} = \rho \mathbf{b} \tag{7.61}$$

or in component form:

$$\rho \frac{\partial^2 u_i}{\partial t^2} - (\nu + \mu) \frac{\partial^2 u_j}{\partial x_i x_j} - \mu \frac{\partial^2 u_i}{\partial x_j x_j} = \rho b_i \quad (i = 1, 2, 3). \tag{7.62}$$

7.4.1.1 Boundary Conditions

The boundary conditions (5.17) in this case become

$$\nu \frac{\partial u_k}{\partial x_k} n_i + \mu \left(\frac{\partial u_i}{\partial x_j} + \frac{\partial u_j}{\partial x_i} \right) n_j = f_i.$$

7.4.1.2 Plane Waves and Propagation Velocity

In the case of no body forces ($\mathbf{b} = 0$), we seek solutions of (7.62) on the form of plane waves:

$$\mathbf{u} = \mathbf{d} \, \exp \left(i \left(\mathbf{x} \cdot \mathbf{n} - V t \right) \right), \tag{7.63}$$

where $\mathbf{n}$ is the constant unit vector normal to the wave front, and the constant vector $\mathbf{d}$ and scalar V need to be determined in order for (7.63) to be a solution of (7.62). Substituting (7.63) into (7.62), we have

$$\left(\rho V^2 - \mu \right) \mathbf{d} = (\nu + \mu)(\mathbf{d} \cdot \mathbf{n}) \mathbf{n}. \tag{7.64}$$

The homogeneous system (7.64) has the following solutions:

$$\mathbf{d} \cdot \mathbf{n} = 0, \qquad V = \pm \sqrt{\frac{\mu}{\rho}} \qquad \text{(transverse waves)}, \tag{7.65}$$

$$\mathbf{d} = \alpha \mathbf{n}, \qquad V = \pm\sqrt{\frac{\nu + 2\mu}{\rho}} \quad \text{(longitudinal waves).} \tag{7.66}$$

Note that in the first case, there are two eigenvectors **d** in the plane orthogonal to **n**, so each transverse wave has a multiplicity of 2. In order for the velocities to make sense, it is necessary to assume that the following inequalities hold for the Lamé constants:

$$\mu > 0, \qquad \nu + 2\mu > 0.$$

Therefore, small free vibrations propagate as a linear combination of six waves: two pairs of transverse waves (the field vibrates in the tangent plane orthogonal to **n***) with velocity* $V = \pm\sqrt{\mu/\rho}$ *and two longitudinal waves (the field oscillates in the normal direction) with velocity* $V = \pm\sqrt{(\nu + 2\mu)/\rho}$.

7.4.1.3 Wave Equation

In the case of the one-dimensional problem where $\mathbf{u} \equiv (u, 0, 0)$ depends only on a spatial variable x, Eq. (7.62) reduces to the classical linear wave equation:

$$\frac{\partial^2 u}{\partial t^2} - c^2 \frac{\partial^2 u}{\partial x^2} = 0, \qquad c = \sqrt{\frac{\nu + 2\mu}{\rho}}, \tag{7.67}$$

which is obviously a particular case of (7.55), since (7.60) in the one-dimensional case becomes

$$W = \rho \frac{c^2}{2} u_x^2.$$

The solution of (7.67) will be discussed later (see Sect. 10.3).

Chapter 8
Fluids

Abstract In this chapter, we will discuss the constitutive class that characterizes fluids, which are bodies for which stress in equilibrium has an isotropic pressure character. The consequences of the entropy principle will be discussed in detail, presenting both EULER's equations for an ideal fluid and the FOURIER–NAVIER–STOKES equations in the case of dissipative fluids. We will also briefly discuss incompressible fluids, BERNOULLI's theorem, and the propagation velocities of plane waves.

8.1 Ideal Fluids and Euler's Equations

Definition 8.1.1 A fluid is said to be ideal if the specific stress at each point is normal to the section and has a pressure character:

$$\mathbf{t}_n = -p_n \mathbf{n}, \quad p_n > 0, \quad \forall\, \mathbf{n}. \tag{8.1}$$

From the validity of CAUCHY's theorem (4.3), we have

$$p_n\, \mathbf{n} = p_1 n_1\, \mathbf{e}_1 + p_2 n_2\, \mathbf{e}_2 + p_3 n_3\, \mathbf{e}_3,$$

and by taking the components of the vectors of both sides, we immediately obtain the result first observed by PASCAL:

$$p_n = p_1 = p_2 = p_3 = p \quad \forall\, \mathbf{n}.$$

In an ideal fluid, the pressure p does not depend on the normal unit vector considered, thus having isotropic pressure: At a given point, the pressure is the same in all directions.

© The Author(s), under exclusive license to Springer Nature Switzerland AG 2024

T. Ruggeri, *Introduction to the Thermomechanics of Continua and Hyperbolic Systems*, La Matematica per il 3+2 167,
https://doi.org/10.1007/978-3-031-69951-1_8

Therefore, we can write

$$\mathbf{t}_n = -p\,\mathbf{n}, \quad p > 0, \quad \forall\,\mathbf{n}, \tag{8.2}$$

which is equivalent to saying that *the stress tensor is isotropic*:

$$\mathbf{t} = -p\mathbf{I}, \quad \mathbf{t} \equiv \begin{pmatrix} -p & 0 & 0 \\ 0 & -p & 0 \\ 0 & 0 & -p \end{pmatrix}, \quad t_{ij} = -p\delta_{ij}. \tag{8.3}$$

For an ideal fluid, thermal conductivity is assumed to be negligible, and thus there is no heat flux:

$$\mathbf{q} = 0. \tag{8.4}$$

By inserting (8.3) and (8.4) into the differential system of balance laws (5.31), we obtain the following *system of balance laws for an ideal fluid*:

$$\begin{cases} \dfrac{\partial \rho}{\partial t} + \dfrac{\partial \rho v_i}{\partial x_i} = 0, \\[2ex] \dfrac{\partial \rho v_j}{\partial t} + \dfrac{\partial}{\partial x_i}\left(\rho v_i v_j + p\delta_{ij}\right) = \rho b_j \quad (j = 1, 2, 3), \\[2ex] \dfrac{\partial}{\partial t}\left(\dfrac{\rho v^2}{2} + \rho e\right) + \dfrac{\partial}{\partial x_i}\left\{\left(\dfrac{\rho v^2}{2} + \rho e + p\right) v_i\right\} = \rho b_j v_j + r. \end{cases} \tag{8.5}$$

The system is not closed since there are five scalar equations for six fields: $\rho(\mathbf{x}, t)$, $\mathbf{v}(\mathbf{x}, t)$, $p(\mathbf{x}, t)$, and $e(\mathbf{x}, t)$. In the usual formulation of fluid thermodynamics, it is preferable to consider the following five unknown fields (corresponding to the five equations): mass density ρ, three components of the velocity vector $\mathbf{v}$, and absolute temperature ϑ. From this perspective, the pressure and internal energy density are considered as constitutive relationships, which, under the assumption of EULER, are local functions of ρ and ϑ:

$$p \equiv p(\rho, \vartheta), \quad e \equiv e(\rho, \vartheta). \tag{8.6}$$

The system (8.5) with the constitutive equations (8.6) is called *Euler's system*.

We will soon see the constraints that arise from the entropy principle for the constitutive equations (8.6).

8.1.1 Boundary Conditions for Ideal Fluids

In the case of an ideal fluid, the boundary condition (5.27) becomes

$$-p\mathbf{n}=\mathbf{f} \quad \forall P \in \Sigma. \tag{8.7}$$

Equation (8.7) states that *the surface of the fluid is arranged in such a way that its outward normal is directed in the direction of the surface force and in the opposite direction.*

This implies, for example, that a fluid inside a container subjected to constant atmospheric pressure will have its free surface horizontally.

8.1.2 Work of Internal Forces in an Ideal Fluid

From (5.22) and (5.23), we have

$$\mathcal{P}^{(i)}=p \operatorname{div} \mathbf{v}.$$

Taking into account (5.4) and (5.69), we obtain the work of internal forces for an ideal fluid:

$$\delta l_*^{(i)}=-\frac{p}{\rho^2}\,\mathrm{d}\rho. \tag{8.8}$$

Sometimes the so-called *specific volume* is defined as

$$V=\frac{1}{\rho}. \tag{8.9}$$

In this case, (8.8) becomes

$$\delta l_*^{(i)}=p\,\mathrm{d}V. \tag{8.10}$$

From (8.10)), we have the interesting result:

In a fluid, the work of internal forces is solely due to volume variations of the fluid. If the ideal fluid is incompressible, the work of internal forces will be zero, similar to what happens for rigid bodies.

8.2 Fourier–Navier–Stokes Dissipative Fluids

Definition 8.2.1 We define a real fluid as a material in which the Cauchy stress tensor has a pressure character only in equilibrium.

Unlike an ideal fluid, the conditions (8.2) and (8.3) hold only in equilibrium for a real fluid. In a nonequilibrium state, the specific stress does not have a normal direction, and we write

$$\mathbf{t} = -p\mathbf{I} + \boldsymbol{\sigma}, \quad t_{ij} = -p\delta_{ij} + \sigma_{ij}, \tag{8.11}$$

where the tensor $\boldsymbol{\sigma} \equiv \left\|\sigma_{ij}\right\| \in Sym$ is called the *viscous stress tensor*, and it is assumed to depend on the symmetric part of the velocity gradient $\mathbf{D}$:

$$\boldsymbol{\sigma} \equiv \boldsymbol{\sigma}(\mathbf{D}), \qquad \boldsymbol{\sigma}(\mathbf{0}) = 0. \tag{8.12}$$

The second condition in (8.12) guarantees that in equilibrium ($\mathbf{v} = 0$ and thus $\mathbf{D} = 0$), there is a coincidence between (8.11) and (8.3).

For a real fluid, heat flux will also be nonzero and will depend on the temperature gradient:

$$\mathbf{q} \equiv \mathbf{q}(\nabla\vartheta), \quad \mathbf{q}(0) = 0. \tag{8.13}$$

The simplest assumption for the constitutive laws (8.12) and (8.13) is to consider a linear relationship. In particular, a linear relationship between $\boldsymbol{\sigma}$ and $\mathbf{D}$ (the *Navier–Stokes hypothesis*) and a linear relationship between $\mathbf{q}$ and $\nabla\vartheta$ (the *Fourier hypothesis*).

Regarding *Fourier's law*, which has already been presented in thermoelasticity, it is written as

$$\mathbf{q} = -\chi \, \nabla\vartheta, \tag{8.14}$$

where the thermal conductivity χ may depend on density and temperature:

$$\chi \equiv \chi(\rho, \vartheta). \tag{8.15}$$

As for the assumption of Navier–Stokes, the linear relationship between $\boldsymbol{\sigma}$ and $\mathbf{D}$ translates into

$$\boldsymbol{\sigma} = \alpha\mathbf{I} + 2\mu\,\mathbf{D},$$

where the scalar μ does not depend on $\mathbf{D}$, while the scalar α must be linear with respect to $\mathbf{D}$. Due to the principle of material indifference, the only linear principal invariant that can be constructed from $\mathbf{D}$ is its trace, which is div$\mathbf{v}$.

It follows that the only satisfying linear relationship (8.12) that satisfies the principle of material indifference is

$$\boldsymbol{\sigma} = \nu \,\mathrm{div}\mathbf{v}\,\mathbf{I} + 2\mu\,\mathbf{D}. \tag{8.16}$$

In terms of components,

$$\sigma_{ij} = \nu \frac{\partial v_k}{\partial x_k}\delta_{ij} + \mu\left(\frac{\partial v_i}{\partial x_j} + \frac{\partial v_j}{\partial x_i}\right).$$

However, it is convenient to represent $\boldsymbol{\sigma}$ as a sum of orthogonal operators. Using the decomposition (1.52) for $\mathbf{D}$

$$\mathbf{D} = \mathbf{D}^D + \frac{1}{3}\mathrm{div}\mathbf{v}\,\mathbf{I},$$

Equation (8.16) takes the form

$$\boldsymbol{\sigma} = \lambda\,\mathrm{div}\mathbf{v}\,\mathbf{I} + 2\mu\mathbf{D}^D, \tag{8.17}$$

where

$$\lambda = \frac{1}{3}(3\nu + 2\mu).$$

Equations (8.17) are known as the *constitutive equations of* NAVIER–STOKES. The scalars λ and μ are called *viscosity coefficients* and can be functions of density and temperature:

$$\lambda \equiv \lambda(\rho,\vartheta), \quad \mu \equiv \mu(\rho,\vartheta). \tag{8.18}$$

From (8.17), we also have that the deviatoric part and the isotropic part of $\boldsymbol{\sigma}$ are, respectively,

$$\boldsymbol{\sigma}^D = 2\mu\mathbf{D}^D, \qquad \boldsymbol{\sigma}^I = \frac{1}{3}\,\mathrm{tr}\boldsymbol{\sigma}\,\mathbf{I} = \lambda\,(\mathrm{div}\,\mathbf{v})\,\mathbf{I}. \tag{8.19}$$

The stress tensor (8.12) for a NAVIER–STOKES fluid becomes

$$\mathbf{t} = -(p+\pi)\,\mathbf{I} + 2\mu\,\mathbf{D}^D, \tag{8.20}$$

where

$$\pi = -\lambda\,\mathrm{div}\mathbf{v}.$$

For this reason, π is called the *dynamic pressure*, while

$$\mathbf{t}^D = \boldsymbol{\sigma}^D = 2\mu\mathbf{D}^D$$

is called the *deviatoric stress tensor*.

A fluid is said to be *Stokesian* or *Stokes fluid* when $\lambda = 0$. In this case, the stress tensor $\boldsymbol{\sigma}$ is purely deviatoric, and the dynamic pressure is identically zero. For this reason, λ is called *bulk viscosity*, while μ is called *shear viscosity*.

The balance equations for a real fluid are the general balance laws (8.21):

$$\begin{cases} \dfrac{\partial \rho}{\partial t} + \dfrac{\partial \rho v_i}{\partial x_i} = 0 \\[2ex] \dfrac{\partial \rho v_j}{\partial t} + \dfrac{\partial}{\partial x_i}\left(\rho v_i v_j + p\delta_{ij} - \sigma_{ij}\right) = \rho b_j \qquad (j = 1, 2, 3) \\[2ex] \dfrac{\partial}{\partial t}\left(\dfrac{\rho v^2}{2} + \rho e\right) + \dfrac{\partial}{\partial x_i}\left\{\left(\dfrac{\rho v^2}{2} + \rho e + p\right) v_i - \sigma_{ij} v_j + q_i\right\} = \rho b_j v_j + r, \end{cases} \tag{8.21}$$

where the viscous stress tensor and heat flux are given by (8.17) and (8.14), respectively. In this system, the density $\rho(\mathbf{x}, t)$, velocity $\mathbf{v}(\mathbf{x}, t)$, and temperature $\vartheta(\mathbf{x}, t)$ are the field variables, and the remaining five quantities need to be specified as constitutive functions:

$$\begin{aligned} &p \equiv p(\rho, \vartheta) \ \ \text{(pressure)}, \ \ e \equiv e(\rho, \vartheta) \ \ \text{(internal energy)}, \\ &\chi \equiv \chi(\rho, \vartheta) \ \ \text{(thermal conductivity)}, \\ &\lambda \equiv \lambda(\rho, \vartheta) \ \ \text{and } \mu \equiv \mu(\rho, \vartheta) \ \ \text{(viscosity coefficients)}. \end{aligned} \tag{8.22}$$

It is worth noting that the case of an ideal fluid, described by Euler's equations, can be considered as a limiting case of the Fourier–Navier–Stokes equations by setting the thermal conductivity and viscosity coefficients to zero.

Under regularity assumptions, the balance equations can be written in terms of the material derivative (3.2):

$$\begin{cases} \dfrac{d\rho}{dt} + \rho \operatorname{div} \mathbf{v} = 0 \\[2ex] \rho \dfrac{dv_j}{dt} + \dfrac{\partial}{\partial x_i}\left(p\delta_{ij} - \sigma_{ij}\right) = \rho b_j \\[2ex] \rho \dfrac{de}{dt} + p \operatorname{div} \mathbf{v} - \boldsymbol{\sigma} \cdot \mathbf{D} + \operatorname{div} \mathbf{q} = r. \end{cases} \tag{8.23}$$

The first equation in (8.23) is immediate. To prove the remaining equations, let us observe that for any function f, thanks to the continuity equation $(8.21)_1$, we have the identity:

$$\frac{\partial \rho f}{\partial t} + \frac{\partial \rho f v_i}{\partial x_i} = \rho \frac{df}{dt}. \tag{8.24}$$

Therefore, from $(8.21)_2$, the second equation in (8.23) is also immediate. As for the third equation, by applying (8.24) to $(8.21)_3$, we have

$$\frac{d\vartheta}{dt} = \frac{1}{\rho \frac{\partial e}{\partial \vartheta}} \left\{ r + \left(\rho^2 \frac{\partial e}{\partial \rho} - p \right) \operatorname{div} \mathbf{v} + \boldsymbol{\sigma} \cdot \mathbf{D} - \operatorname{div} \mathbf{q} \right\}. \tag{8.25}$$

In the next section, we will see the restrictions imposed by the principle of entropy on physically acceptable constitutive equations (8.22). These restrictions also hold as limiting cases for Euler's equations.

8.3 Entropy Principle for a Fluid

Now we want to deduce the constraints that arise from the entropy principle in the case of a Fourier–Navier–Stokes fluid proving the following theorem:

Theorem 8.3.1 *The entropy principle in the case of a Fourier–Navier–Stokes fluid requires the existence of a free energy function ψ, dependent on ρ and ϑ. The admissible constitutive functions $p \equiv p(\rho, \vartheta)$, $e \equiv e(\rho, \vartheta)$, $S \equiv S(\rho, \vartheta)$ are given by*

$$p = \rho^2 \frac{\partial \psi}{\partial \rho}, \qquad S = -\frac{\partial \psi}{\partial \vartheta}, \qquad e = \psi - \vartheta \frac{\partial \psi}{\partial \vartheta}. \tag{8.26}$$

The remaining constitutive functions λ, μ, and χ must be nonnegative

$$\lambda(\rho, \vartheta) \geq 0, \qquad \mu(\rho, \vartheta) \geq 0, \qquad \chi(\rho, \vartheta) \geq 0. \tag{8.27}$$

Proof Let us rewrite the entropy principle (6.7) using the material derivative and taking into account $(8.23)_1$:

$$\rho \frac{dS}{dt} + \frac{\partial}{\partial x_i} \left(\frac{q_i}{\vartheta} \right) - \frac{r}{\vartheta} = \Sigma \geq 0. \tag{8.28}$$

Assuming that the entropy density also depends on density and temperature, i.e.,

$$S \equiv S(\rho, \vartheta),$$

we have, from (8.28), $(8.23)_1$, and (8.25),

$$\left\{\frac{1}{\vartheta}-\frac{\frac{\partial S}{\partial \vartheta}}{\frac{\partial e}{\partial \vartheta}}\right\}(\operatorname{div}\mathbf{q}-r)+\operatorname{div}\mathbf{v}\left\{-\rho^2\frac{\partial S}{\partial \rho}+\left(\rho^2\frac{\partial e}{\partial \rho}-p\right)\frac{\frac{\partial S}{\partial \vartheta}}{\frac{\partial e}{\partial \vartheta}}\right\}$$
$$+\frac{\frac{\partial S}{\partial \vartheta}}{\frac{\partial e}{\partial \vartheta}}\boldsymbol{\sigma}\cdot\mathbf{D}-\frac{1}{\vartheta^2}\mathbf{q}\cdot\nabla\vartheta=\Sigma\geq 0. \tag{8.29}$$

Note that in (8.29), the terms in curly braces are functions only on ρ and ϑ and they are multiplied by arbitrary linear spatial first derivatives ($\operatorname{div}\mathbf{q}$ and $\operatorname{div}\mathbf{v}$).

The remaining two terms, on the other hand, are quadratic in the spatial derivatives according to (8.17) and (8.14). Therefore, in order for the inequality (8.29) to hold for any possible process, the terms in curly braces must necessarily be zero:

$$\begin{cases}\vartheta\dfrac{\partial S}{\partial \vartheta}=\dfrac{\partial e}{\partial \vartheta}\\[2ex] \vartheta\dfrac{\partial S}{\partial \rho}=\dfrac{\partial e}{\partial \rho}-\dfrac{p}{\rho^2}.\end{cases} \tag{8.30}$$

The equations (8.30) are equivalent to the differential form of Gibbs equation:

$$\vartheta dS=de-\frac{p}{\rho^2}d\rho. \tag{8.31}$$

We have seen in (8.8) that for an ideal fluid and for a real fluid in equilibrium,

$$\delta l_*^{(i)}=-\frac{p}{\rho^2}d\rho.$$

Therefore, the GIBBS formula, as in the elastic case (7.26), becomes

$$\vartheta dS=de+\delta l_*^{(i)}.$$

The GIBBS equation is said to characterize a *local equilibrium* and expresses the fact that part of the internal energy is transformed into heat ϑdS, while the rest is transformed into work by internal forces. By introducing the HELMHOLTZ *free energy* as

$$\psi=e-\vartheta S \tag{8.32}$$

and taking into account (8.31), we have

$$d\psi=\frac{p}{\rho^2}d\rho-Sd\vartheta. \tag{8.33}$$

From (8.32) and (8.33), we obtain (8.26).

Substituting into (8.29), (8.17), (8.14), and $(8.30)_1$ and taking into account the decomposition (1.52), we have

$$\Sigma = \frac{1}{\vartheta}\left(\lambda \operatorname{div}\mathbf{v}\,\mathbf{I} + 2\mu\mathbf{D}^D\right)\cdot\mathbf{D} + \frac{\chi}{\vartheta^2}|\nabla\vartheta|^2 =$$

$$\frac{1}{\vartheta}\left(\lambda\,(\operatorname{div}\mathbf{v})^2 + 2\mu\|\mathbf{D}^D\|^2\right) + \frac{\chi}{\vartheta^2}|\nabla\vartheta|^2 \geq 0.$$

Since the three quadratic terms are independent of one another, we necessarily have the inequalities (8.27). □

8.4 Some Special Cases of Fluids

In this section, we will examine some special cases of fluids and provide some simple examples of solutions.

8.4.1 Perfect Gases

Compressible fluids are usually gases. Among them, we have the so-called *perfect gases*, for which the free energy is given by

$$\psi = \frac{k\vartheta}{m}\ln\left(\frac{\alpha\rho}{\vartheta^{\frac{1}{\gamma-1}}}\right),$$

where the constants k, m, and γ represent, respectively, the Boltzmann constant, the molecular weight, and the ratio of specific heats ($\gamma = 5/3$ for a monatomic gas and $\gamma = 7/5$ for a diatomic gas), while α is an arbitrary integration constant.

From $(8.26)_1$, we immediately obtain the well-known formula for pressure in the case of a perfect gas:

$$p = \frac{k}{m}\rho\vartheta. \tag{8.34}$$

Furthermore, the internal energy and entropy density can be obtained from $(8.26)_{2,3}$:

$$e = \frac{k\vartheta}{m(\gamma-1)}; \qquad S = \frac{k}{m}\left\{\frac{1}{\gamma-1} + \log\left(\frac{\vartheta^{\frac{1}{\gamma-1}}}{\alpha\rho}\right)\right\}. \tag{8.35}$$

It can be observed that in a perfect gas, the internal energy depends only on temperature. From (8.34) and (8.35), the relationships between internal energy and entropy with pressure and density can be deduced[1]:

$$e = \frac{p}{\rho\,(\gamma - 1)}, \qquad S = c_V \log\left(\frac{p}{\rho^\gamma}\right) + S_0 \tag{8.36}$$

(where S_0 is an arbitrary constant related to α). Sometimes it is convenient to express the constitutive equations using the thermodynamic variables (ρ, S) as independent variables instead of (ρ, ϑ). In this case, from (8.36), we immediately have

$$p = A(S)\,\rho^\gamma; \qquad e = \frac{A(S)}{\gamma - 1}\,\rho^{\gamma-1}, \tag{8.37}$$

where

$$A(S) = \exp\left(\frac{S - S_0}{c_V}\right).$$

8.4.1.1 Static Solution of a Gas in the Presence of Gravity

Let us consider a compressible fluid at rest ($\mathbf{v} = 0$) and subject to the force of gravity, and let us look for a static solution (independent of time) of (8.23) with (8.17) and (8.14).

Assuming the x_3 axis to be vertical and pointing upward, we have $b_j = -g\delta_{j3}$, where g is the gravity acceleration. Therefore, the system (8.23), taking into account (8.34) and (8.35), becomes

$$\begin{cases} \dfrac{k}{m}\dfrac{\partial \rho\vartheta}{\partial x_i} = -\rho g \delta_{i3} \\[2ex] \operatorname{div}(\chi \nabla \vartheta) = -r. \end{cases} \tag{8.38}$$

This problem arises, for example, in the case of air when one wants to understand how density and temperature vary with increasing altitude. If we consider heat conductivity constant and we neglect the radiative effects of the Earth on the air and set $r = 0$, then there exits the particular solution of (8.38):

$$\begin{cases} \vartheta = \vartheta_0 \\[2ex] \rho = \rho_0 \exp\left(-\dfrac{gm}{k\vartheta}\,x_3\right), \end{cases} \tag{8.39}$$

[1] It should be noted that $R = k/m = c_p - c_V$ and $\gamma = c_p/c_V$, where c_p and c_V represent the specific heats at constant pressure and volume, respectively.

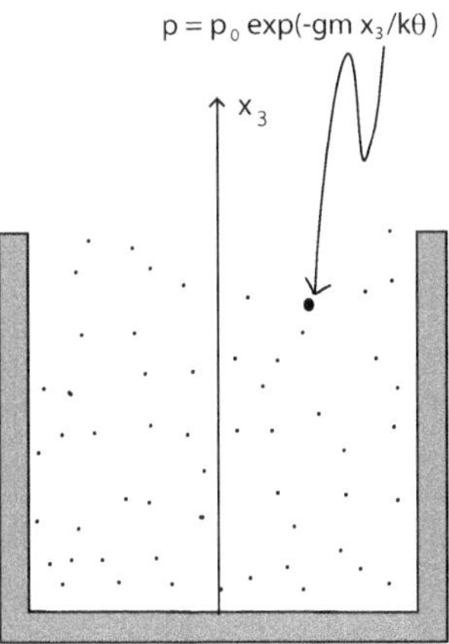

Fig. 8.1 Variation of pressure with altitude

where ϑ_0 and ρ_0 are the temperature and density, respectively, at $x_3 = 0$. *From* (8.39), *we can see that the density decreases exponentially with increasing altitude. From* (8.34) *and* (8.39), *we can also see that the pressure decreases with the same exponential law* (see Fig. 8.1):

$$p = p_0 \exp\left(-\frac{gm}{k\vartheta}\, x_3\right).$$

It should be noted that the solution (8.39) is also a solution of the Euler system.

8.4.2 Incompressible Navier–Stokes Fluids

Liquids are fluids that, to a good approximation, do not change their volume and are therefore *incompressible*. Considering isothermal processes and using (2.34) and (2.35), we have

$$J = 1 \quad \Leftrightarrow \quad \rho = \rho^* = const.$$

Taking into account (8.17), the mass and momentum balance laws become

$$\begin{cases} \dfrac{\partial v_i}{\partial x_i} = 0 \\[2ex] \rho^* \left(\dfrac{\partial v_j}{\partial t} + \dfrac{\partial v_j}{\partial x_i} v_i\right) + \dfrac{\partial p}{\partial x_j} - \mu \Delta v_j = \rho^* b_j \end{cases} \tag{8.40}$$

(where Δ represents the Laplacian operator: $\Delta = \partial^2/\partial x_i \partial x_i$).

For a given ρ^* and μ, the system is closed as there are four equations for the four unknowns v_j $(j = 1, 2, 3)$ and p.

Note that p is now an unknown, as it plays the role of a reaction corresponding to the incompressibility constraint.

Taking into account (8.20), the boundary conditions (5.17) in this case become

$$-p n_i + \mu \left(\frac{\partial v_i}{\partial x_j} + \frac{\partial v_j}{\partial x_i} \right) n_j = f_i .$$

8.4.3 Compressible Euler Fluids

We have already seen that an ideal fluid can be seen as a special case of a real fluid by neglecting thermal conductivity and viscosity coefficients: $\chi = \lambda = \mu = 0$. However, the restrictions (8.26) of the entropy principle still hold:

In an Euler fluid, only one constitutive function needs to be known, the free energy $\psi(\rho, \vartheta)$ *from which pressure, internal energy, and entropy density are derived using (8.26).*

In particular, for a perfect fluid, (8.34), (8.35), and (8.36) hold. The equations (8.23) become in this case

$$\begin{cases} \dfrac{d\rho}{dt} + \rho \operatorname{div} \mathbf{v} = 0 \\[2ex] \rho \dfrac{d\mathbf{v}}{dt} + \operatorname{grad} p = \rho \mathbf{b} \\[2ex] \rho \dfrac{de}{dt} + p \operatorname{div} \mathbf{v} = r. \end{cases} \tag{8.41}$$

Every regular solution of (8.41) is also a solution of the entropy law (8.28), which becomes

$$\rho \frac{dS}{dt} - \frac{r}{\vartheta} = 0. \tag{8.42}$$

We observe that in this ideal case the production of entropy $\Sigma = 0$ and the process is reversible. Moreover if there is no external heat production, $r = 0$, it follows by (8.42) that the entropy remains constant along the fluid particle motion.

Sometimes it is convenient to work with the fields $(\rho, \mathbf{v}, S)$ instead of $(\rho, \mathbf{v}, \vartheta)$ and replace $(8.41)_3$ with (8.42), obtaining the system:

$$\begin{cases} \dfrac{d\rho}{dt} + \rho \operatorname{div} \mathbf{v} = 0 \\ \rho \dfrac{d\mathbf{v}}{dt} + \operatorname{grad} p = \rho \mathbf{b} \\ \rho \dfrac{dS}{dt} = \dfrac{r}{\vartheta}, \end{cases} \tag{8.43}$$

where now p is a constitutive function of ρ and S:

$$p \equiv p(\rho, S). \tag{8.44}$$

8.4.3.1 Linearized Equations and Speed of Sound

Now let us consider the case where there are no mass forces $\mathbf{b} = 0$ and no heat sources $r = 0$. In this case, there exists a constant solution of (8.43):

$$\rho = \rho_0, \qquad S = S_0, \qquad \mathbf{v} = \mathbf{v}_0.$$

Let us consider small perturbations from this equilibrium state:

$$\rho = \rho_0 + \tilde{\rho}, \qquad \mathbf{v} = \mathbf{v}_0 + \tilde{\mathbf{v}}, \qquad S = S_0 + \tilde{S},$$

where the quantities with tildes are assumed to be small (for stability reasons), and we will linearize the equations (8.43) around the state $(\rho_0, \mathbf{v}_0, S_0)$ by neglecting the nonlinear terms in $\tilde{\rho}, \tilde{\mathbf{v}}, \tilde{S}$ and their partial derivatives. After some simple calculations, we obtain:

$$\begin{cases} \dfrac{\partial \tilde{\rho}}{\partial t} + v_{0i} \dfrac{\partial \tilde{\rho}}{\partial x_i} + \rho_0 \operatorname{div} \tilde{\mathbf{v}} = 0 \\ \rho_0 \left(\dfrac{\partial \tilde{v}_j}{\partial t} + v_{0i} \dfrac{\partial \tilde{v}_j}{\partial x_i} \right) + \dfrac{\partial \tilde{p}}{\partial x_j} = 0 \\ \rho_0 \left(\dfrac{\partial \tilde{S}}{\partial t} + v_{0i} \dfrac{\partial \tilde{S}}{\partial x_i} \right) = 0 \,. \end{cases} \tag{8.45}$$

Let us look for a plane wave solution of (8.45), omitting the tildes from now on:

$$
\begin{aligned}
\rho &= \delta\rho \exp\{i(\mathbf{k}\cdot\mathbf{x} - \omega t)\} \\
v_j &= \delta v_j \exp\{i(\mathbf{k}\cdot\mathbf{x} - \omega t)\} \\
S &= \delta S \exp\{i(\mathbf{k}\cdot\mathbf{x} - \omega t)\},
\end{aligned} \tag{8.46}
$$

where $\delta\rho, \delta v_j$, and δS are the (small) constant amplitudes of the wave, $\mathbf{k} \equiv (k_1, k_2, k_3)$, and ω are the *wave vector* and the so-called *frequency*, which are both constant. Let $\mathbf{n}$ be the unit vector normal to the wavefront such that

$$\mathbf{k} = k\mathbf{n}; \quad k \text{ is the wave number},$$

and let us define the so-called *phase velocity*

$$v_f = \frac{\omega}{k}.$$

Taking into account (8.44), we have

$$\frac{\partial p}{\partial x_i} = p_\rho \frac{\partial \rho}{\partial x_i} + p_S \frac{\partial S}{\partial x_i},$$

where we denote

$$p_\rho = \left(\frac{\partial p}{\partial \rho}\right)_S, \quad p_S = \left(\frac{\partial p}{\partial S}\right)_\rho. \tag{8.47}$$

Substituting (8.46) into (8.45), we find after some simple calculations that the amplitudes must satisfy the following homogeneous linear system:

$$
\begin{cases}
(-v_f + v_{0n})\delta\rho + \rho_0 \delta\, v_n = 0 \\
\rho_0(-v_f + v_{0n})\delta v_j + n_j\{(p_\rho)_0 \delta\rho + (p_S)_0 \delta S\} = 0 \\
(-v_f + v_{0n})\delta S = 0,
\end{cases} \tag{8.48}
$$

where we have defined $\delta v_n = \delta\mathbf{v}\cdot\mathbf{n}$ and $v_{0n} = \mathbf{v}_0 \cdot \mathbf{n}$, and the subscript 0 indicates that the derivatives p_ρ and p_S are to be evaluated at $\rho = \rho_0$ and $S = S_0$.

From (8.48), we immediately see that there are two different categories of waves:

Material Waves

$$v_f = v_{0n},$$

with $\delta v_n = 0$; $\delta p = (p_\rho)_0 \delta\rho + (p_S)_0 \delta S = 0$ and δS arbitrary.
$v_f = v_{0n}$ is a triple root of the homogeneous system (8.48). In this case, the wave propagates with a phase velocity equal to the normal component of the unperturbed velocity. The vector $\delta\mathbf{v}$ is tangent to the plane of the wave (*transverse wave*), and the wave does not carry pressure perturbations.

Sound Waves

$$v_f \neq v_{0n};$$

then from $(8.48)_3$, we have $\delta S = 0$ and from $(8.48)_2$, we deduce that $\delta\mathbf{v}$ is parallel to $\mathbf{n}$ (*longitudinal wave*). By taking the scalar product of $(8.48)_2$ with $\mathbf{n}$, we obtain the homogeneous system in $\delta\rho$ and δv_n:

$$(-v_f + v_{0n})\delta\rho + \rho_0 \delta v_n = 0,$$

$$c_0^2 \delta\rho + \rho_0(-v_f + v_{0n})\delta v_n = 0,$$

where

$$c = \sqrt{p_\rho} \text{ (sound velocity).} \tag{8.49}$$

By solving this system, we find that there are two waves moving relative to the fluid with velocities $-c_0$ and c_0. These *sound waves* are characterized by variations in density and pressure and are longitudinal. It is important to note that in the expression for c given by (8.49), we need that

$$p_\rho = \left(\frac{\partial p}{\partial \rho}\right)_S > 0,$$

which implies that the pressure increases with increasing density. For a perfect gas, we have from (8.37) and from (8.34)

$$\left(\frac{\partial p}{\partial \rho}\right)_S = \gamma \frac{p}{\rho},$$

where γ is the ratio of specific heat, and we can express the speed of sound as

$$c = \sqrt{\gamma \frac{p}{\rho}} = \sqrt{\gamma \frac{k}{m} \vartheta},$$

where ϑ is the temperature. This shows that *the speed of sound in a perfect gas depends only on the temperature and increases with it.*

In summary, material waves are transverse waves that do not carry pressure perturbations, while sound waves are longitudinal waves characterized by variations in density and pressure and propagate with the speed of sound, which depends on the properties of the medium (such as the unperturbed temperature in the case of a perfect gas).

8.4.4 Incompressible Euler Ideal Fluids and Bernoulli's Theorem

For an incompressible EULER fluid, the equations (8.40) simplify to

$$\begin{cases} \operatorname{div} \mathbf{v} = 0 \\ \rho^* \dfrac{d\mathbf{v}}{dt} + \operatorname{grad} p = \rho^* \mathbf{b}. \end{cases} \tag{8.50}$$

Consider an incompressible fluid subjected to the gravity ($b_j = -g\delta_{j3}$), with the axes chosen such that x_3 is vertical and oriented upward. In this case, the equations (8.50) become

$$\begin{cases} \operatorname{div} \mathbf{v} = 0 \\ \rho^* \dfrac{dv_j}{dt} + \dfrac{\partial p}{\partial x_j} = -\rho^* g \delta_{j3}. \end{cases} \tag{8.51}$$

Now, let us consider a solution of (8.51) corresponding to a *steady flow*, i.e., a process where physical quantities do not explicitly depend on time. This implies that for any function $f(\mathbf{x}(t), t)$,

$$\frac{\partial f}{\partial t} = 0 \qquad \text{and} \qquad \frac{df}{dt} = \frac{\partial f}{\partial x_i} v_i, \tag{8.52}$$

meaning that the material time derivative is solely due to the motion of the particles.

Multiplying $(8.51)_2$ by v_j and summing over j, we have

$$\frac{d}{dt}\left(\rho^* \frac{v^2}{2}\right) + v_j \frac{\partial p}{\partial x_j} = -\rho^* g v_3.$$

Taking (8.52) into account and using

$$v_3 = \frac{dx_3}{dt},$$

we obtain:

$$\frac{d}{dt}\left(\rho^*\frac{v^2}{2}+p+\rho^* g x_3\right)=0,$$

which means the term in parentheses is independent of time. Then

$$\frac{v^2}{2g}+\frac{p}{\rho^* g}+x_3=\text{constant}. \tag{8.53}$$

During steady motion, the sum of these three terms remains constant.

Since each of them has the dimensions of length, the theorem due to BERNOULLI is called the *theorem of the three heights*, as follows:

- x_3 represents the *geometric height*.
- $\frac{v^2}{2g}$ represents the *kinetic height* (the height at which a material point thrown downward with zero initial velocity reaches the velocity v).
- $\frac{p}{\rho^* g}$ is called the *piezometric height*.

In particular, if the fluid is in equilibrium with $v=0$, from (8.53) we have

$$p=p_0-x_3\rho_* g, \tag{8.54}$$

where p_0 denotes the atmospheric pressure at the free surface of the fluid with $x_3=0$.

From (8.54), it is well known that the pressure in an incompressible fluid in equilibrium increases linearly with depth (remember that, as indicated in Fig. 8.2, the x_3 axis is directed upward).

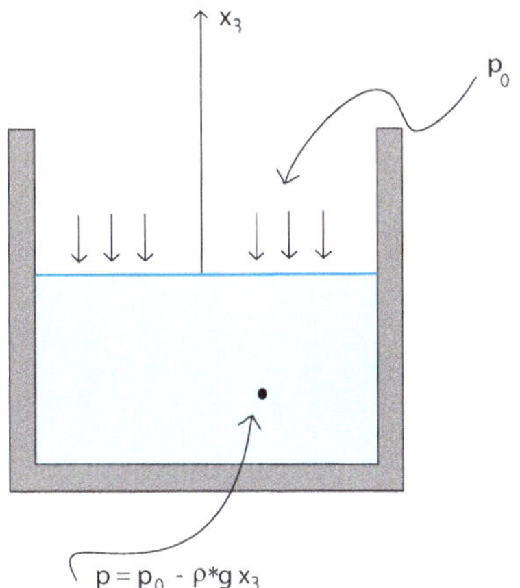

Fig. 8.2 Variation of pressure in a liquid with depth

8.5 Fluid Equations in Lagrangian Variables

It has been seen that in the case of nonlinear thermoelasticity, it is necessary to use the Lagrangian formulation, while for fluids we use Eulerian framework. However, it is always possible, and sometimes even convenient, to rewrite the differential system of fluids using Lagrangian variables. In the case of an ideal fluid of EULER, the system coincides with the general one (5.59) where the stress tensor given by (8.2) is substituted. In this case, the first PIOLA–KIRCHHOFF tensor is given by

$$T_{iA} = -p\,(\rho, \vartheta)\, F^C_{iA} \qquad \Longleftrightarrow \qquad \mathbf{T} = -p\,(\rho, \vartheta)\, \mathbf{F}^C,$$

and therefore (5.59) becomes

$$\begin{cases} \rho = \rho^*/J \\[2ex] \dfrac{\partial \rho^* v_i}{\partial t} + \dfrac{\partial p(\rho, \vartheta) F^C_{iA}}{\partial X_A} = \rho^* b_i \\[2ex] \dfrac{\partial}{\partial t}\left(\rho^* \dfrac{v^2}{2} + \rho^* e(\rho, \vartheta)\right) + \dfrac{\partial}{\partial X_A}\left(p(\rho, \vartheta) F^C_{iA} v_i\right) = \rho^* b_i v_i + r^*. \end{cases} \tag{8.55}$$

In the one-dimensional case where it is assumed that the quantities depend only on $X = X_1$, recalling (7.54) and (8.55) becomes (in the absence of source terms)

$$\begin{cases} \rho = \rho^*/J \\[2ex] \dfrac{\partial \rho^* v}{\partial t} + \dfrac{\partial p(\rho, \vartheta)}{\partial X} = 0 \\[2ex] \dfrac{\partial}{\partial t}\left(\rho^* \dfrac{v^2}{2} + \rho^* e(\rho, \vartheta)\right) + \dfrac{\partial}{\partial X}\left(p(\rho, \vartheta) v\right) = 0. \end{cases} \tag{8.56}$$

Recalling that in the present one-dimensional case

$$u_X = F - 1 = J - 1 = \frac{\rho^*}{\rho} - 1,$$

it is convenient to rewrite the second-order system (8.56) as a first-order system by introducing the variable $w = u_X$:

$$\begin{cases} \rho^* \dfrac{\partial v}{\partial t} + \dfrac{\partial p(w, \vartheta)}{\partial X} = 0 \\ \\ \dfrac{\partial w}{\partial t} - \dfrac{\partial v}{\partial X} = 0 \\ \\ \rho^* \dfrac{\partial}{\partial t} \left(\dfrac{v^2}{2} + e(w, \vartheta) \right) + \dfrac{\partial}{\partial X} (p(w, \vartheta) v) = 0, \end{cases}$$

in the unknowns (v, w, ϑ).

If we consider the purely mechanical case, neglect the dependence on temperature, and set the constant $\rho^* = 1$, the previous system becomes

$$\begin{cases} \dfrac{\partial v}{\partial t} + \dfrac{\partial p(w, \vartheta)}{\partial X} = 0 \\ \\ \dfrac{\partial w}{\partial t} - \dfrac{\partial v}{\partial X} = 0. \end{cases}$$

This system is called the *p-system* and is widely used in the literature of hyperbolic partial differential equations as a simple fluid-dynamic model.

Chapter 9
Rigid Heat Conductor

Abstract In this brief chapter, we consider the case of a rigid body that conducts heat and derive the classical *Fourier heat equation*. We then discuss the so-called *heat paradox* which arises from the Fourier equation, where heat propagates with infinite velocity. We present the alternative model proposed by CATTANEO to resolve this paradox.

9.1 Heat Equation

For a rigid body that conducts heat with constant density ρ^*, the only balance law (5.31) that needs to be satisfied is the energy balance, which takes the form:

$$\rho^* \frac{\partial e}{\partial t} + \operatorname{div} \mathbf{q} = r. \tag{9.1}$$

The classical constitutive equation associated with Eq. (9.1) is given by FOURIER's law:

$$\mathbf{q} = -\chi \nabla \vartheta, \tag{9.2}$$

where we assume

$$e \equiv e(\vartheta) \quad \text{and} \quad \chi \equiv \chi(\vartheta). \tag{9.3}$$

By substituting equations (9.2) and (9.3) into Eq. (9.1), we obtain

$$\rho^* c_V(\vartheta) \frac{\partial \vartheta}{\partial t} - \operatorname{div}(\chi \nabla \vartheta) = r, \tag{9.4}$$

© The Author(s), under exclusive license to Springer Nature Switzerland AG 2024

T. Ruggeri, *Introduction to the Thermomechanics of Continua and Hyperbolic Systems*, La Matematica per il 3+2 167,
https://doi.org/10.1007/978-3-031-69951-1_9

where we have indicated

$$c_V(\vartheta) = e'(\vartheta) = \frac{de}{d\vartheta} \quad \text{(specific heat at constant volume).}$$

Equation (9.4) can be rewritten as

$$\frac{\partial \vartheta}{\partial t} - \mu\, \Delta\vartheta - \nu\,(\text{grad}\,\vartheta)^2 = r, \tag{9.5}$$

where we have defined

$$\mu(\vartheta) = \frac{\chi(\vartheta)}{\rho^* c_V(\vartheta)}; \quad \nu(\vartheta) = \frac{\chi'(\vartheta)}{\rho^* c_V(\vartheta)}; \quad \chi'(\vartheta) = \frac{d\chi(\vartheta)}{d\vartheta}.$$

Equation (9.5) is known as the *heat equation* and governs the temperature evolution in the rigid conductor.

In the simple case where both χ and c_V are constants and $r = 0$, (9.5) takes the classical form of heat equation:

$$\frac{\partial \vartheta}{\partial t} - \mu\, \Delta\vartheta = 0. \tag{9.6}$$

The differential equation (9.6) is a typical equation of *diffusion*, and mathematically, it is a *parabolic* differential equation (see the next chapter).

If we assign the initial temperature as a function of position:

$$\vartheta(\mathbf{x}, 0) = \vartheta_0(\mathbf{x}), \tag{9.7}$$

it can be verified that the solution of (9.6) satisfying (9.7) is [18]

$$\vartheta(\mathbf{x}, t) = \frac{1}{(4\pi\mu t)^{d/2}} \int_{-\infty}^{+\infty} \vartheta_0(\mathbf{y}) \exp\left(-\frac{(\mathbf{y}-\mathbf{x})^2}{4\mu t}\right) d\mathbf{y}, \tag{9.8}$$

where d is the space dimension. In the case of $d = 3$, one characteristic of (9.8) is that $\vartheta(\mathbf{x}, t)$ is nonzero for all $\mathbf{x} \in \mathbb{R}^3$ after $t > 0$, even if $\vartheta_0(\mathbf{x})$ is nonzero only in a limited region $\mathcal{D} \subset \mathbb{R}^3$. This phenomenon is known as the *heat paradox*: A localized disturbance at $t = 0$ in $\mathcal{D}$ *instantaneously propagates* throughout space![1]

This is a typical phenomenon of parabolic-type equations, and to eliminate the paradox, it is necessary to modify the model to make the differential equation for the temperature of *hyperbolic* type.

[1] Note that in the case of $r = 0$, Eq. (9.6) can be considered as the linearized equation of (9.5) and ϑ represents the temperature perturbation with respect to a constant temperature.

9.2 Cattaneo Equation

CATTANEO observed that in nonequilibrium conditions, the temperature changes over time. Therefore, he hypothesized that the heat flux depends not only on the temperature gradient but also on the gradient of the temperature derivative:

$$\mathbf{q} = -\chi \nabla \vartheta + \chi \tau \nabla \dot{\vartheta} \qquad (\dot{\vartheta} = \partial \vartheta / \partial t), \tag{9.9}$$

where the coefficient τ has the dimensions of time.

Equation (9.9) can also be written as

$$\mathbf{q} = -\chi \left(1 - \tau \frac{\partial}{\partial t}\right) \nabla \vartheta. \tag{9.10}$$

If τ is very small, the inverse operator can be approximated as

$$\left(1 - \tau \frac{\partial}{\partial t}\right)^{-1} \simeq 1 + \tau \frac{\partial}{\partial t}. \tag{9.11}$$

From Eqs. (9.11) and (9.10), we have

$$\left(1 + \tau \frac{\partial}{\partial t}\right) \mathbf{q} = -\chi \nabla \vartheta.$$

Thus, we obtain

$$\tau \frac{\partial \mathbf{q}}{\partial t} + \mathbf{q} = -\chi \nabla \vartheta, \tag{9.12}$$

One noticeable difference between Eq. (9.12) and the classical Fourier equation (9.2) is observed when ϑ is constant. In the case of Fourier's law (9.2), the heat flux $\mathbf{q}$ is identically zero, whereas in the case of Cattaneo's equation (9.12), we have

$$\mathbf{q}(t) = \mathbf{q}(0)\, \mathrm{e}^{-t/\tau},$$

which means that it takes some time for $\mathbf{q}$ to become negligible. This is why τ is called *relaxation time*.

In general, τ can also be a function of ϑ. In the simple case where τ, c_V, and χ are constants and $r = 0$, differentiating equation (9.1) with respect to time and applying the divergence operator to Eq. (9.12), we obtain the equation:

$$\tau \frac{\partial^2 \vartheta}{\partial t^2} + \frac{\partial \vartheta}{\partial t} - \mu \Delta \vartheta = 0. \tag{9.13}$$

As we will see in the subsequent Sect. (10.2.2), this is a hyperbolic-type equation and is known in the literature as the *telegrapher's equation*. The heat paradox is resolved when $\tau > 0$. When $\tau \to 0$, Eq. (9.13) converges to the classical heat equation (9.6).

Chapter 10
Hyperbolic Systems

Abstract In this chapter, we will provide a brief overview of the nature of the differential systems we have presented, focusing on the study of the class of equations known as *hyperbolic*. This class includes the Euler equations, thermoelasticity equations (when thermal conductivity is negligible), and the system governing heat propagation with the Cattaneo equation. Later, we will introduce the method of characteristics to determine the solution of the wave equation in one spatial dimension and, more generally, for a generic linear homogeneous hyperbolic system. As an example of a nonlinear hyperbolic equation, we will present the *Burgers' equation*, showing that regular solutions exist only for a finite time (*critical time*).

10.1 Classification

We have seen that when constitutive equations are assigned, the system of balance laws becomes a balanced system in which the number of unknowns is equal to the number of equations. The resulting system is a *system of partial differential equations*. The theory of partial differential equations is extremely complex, and there is no entirely general classification. However, there are three general classes of systems: *elliptic*, *parabolic*, and *hyperbolic* systems.

The first class usually governs *equilibrium* phenomena, while the latter two are typical of *evolution problems*. Specifically, parabolic systems are associated with *diffusion* problems, while hyperbolic systems are typical of *propagation* phenomena. Here, we aim to define hyperbolic systems and consider parabolic and elliptic systems as limiting or degenerate cases of hyperbolic systems.

We have already noted that the mathematical structure of the balance laws corresponds to the divergence in space-time of vectors in $\mathbb{R}^N$ (see (5.32) and (5.34)). Let $\mathbf{u} \in \mathbb{R}^N$ represent the vector of unknown fields, and consider locally constitutive

© The Author(s), under exclusive license to Springer Nature Switzerland AG 2024
T. Ruggeri, *Introduction to the Thermomechanics of Continua and Hyperbolic Systems*, La Matematica per il 3+2 167,
https://doi.org/10.1007/978-3-031-69951-1_10

equations. In other words, we can rewrite (5.34) as follows:[1]

$$\frac{\partial \mathbf{F}^{\alpha}(\mathbf{u})}{\partial x^{\alpha}} = \mathbf{f}(\mathbf{u}). \tag{10.1}$$

Here, $\mathbf{F}^{\alpha}$ and $\mathbf{f}$ are local functions of $\mathbf{u}$. As we will see, this class includes the Euler equations of fluid dynamics, the Cattaneo heat equation for rigid heat conductors, and the equations of thermoelasticity.

Equation (10.1) can be written in the equivalent form:

$$\mathbf{A}^{\alpha}(\mathbf{u})\frac{\partial \mathbf{u}}{\partial x^{\alpha}} = \mathbf{f}(\mathbf{u}), \tag{10.2}$$

where the matrices $\mathbf{A}^{\alpha}$ $(N \times N)$ are given by

$$\mathbf{A}^{\alpha} = \frac{\partial \mathbf{F}^{\alpha}}{\partial \mathbf{u}}, \qquad (\alpha = 0, 1, 2, 3). \tag{10.3}$$

In relation to the system written in this form, we have:

(i) *Linear system*: The matrices $\mathbf{A}^{\alpha}$ are constants, and $\mathbf{f}$ is linear in $\mathbf{u}$.
(ii) *Semi-linear system*: The matrices $\mathbf{A}^{\alpha}$ are constants, but $\mathbf{f}$ is nonlinear.
(iii) *Quasi-linear*: The $\mathbf{A}^{\alpha}$ are functions of the field $\mathbf{u}$, but the system (10.2) is linear in the first derivatives (highest-order derivatives).

Definition 10.1.1 (Hyperbolic System) *A system of the type (10.2) is said to be hyperbolic in the temporal direction t if:*

(a) $\det \mathbf{A}^0 \neq 0$.
(b) The following generalized eigenvalue problem

$$(\mathbf{A}_n - \lambda \mathbf{A}^0)\mathbf{d} = 0, \quad (\mathbf{A}_n = \mathbf{A}^i n_i) \tag{10.4}$$

has all real eigenvalues λ, *and the eigenvectors* $\mathbf{d}$ *form a basis of the* N*-dimensional vector space (i.e., they are linearly independent) for all* $\mathbf{n} \in \mathbb{R}^3$: $\|\mathbf{n}\| = 1$. *The* λ *are called characteristic velocities.*

Note that in the linear and semi-linear cases, the λ values are all constant, while in the quasi-linear case, they depend on the field $\mathbf{u}$. This implies that in the nonlinear case, hyperbolicity may be lost for certain field values.

[1] The repeated Greek indices, such as α, denote summation from 0 to 3.

To understand the physical meaning of the hyperbolicity definition, we will prove later that for linear systems:

(1) The eigenvalues λ physically represent the velocities of the fronts (waves) that have unit normal $\mathbf{n}$. The reality condition associated with λ implies that propagation can indeed occur.[2]
(2) $\det \mathbf{A}^0 \neq 0$ excludes the possibility of any eigenvalue λ tending to infinity (it can be seen immediately that if $\det \mathbf{A}^0 \to 0$, then at least one $\lambda \to \infty$). Essentially, this corresponds to behavior of *parabolic* type.
(3) The fact that all eigenvectors $\mathbf{d}$ are linearly independent implies that an arbitrary Cauchy initial condition can be decomposed into N *modes*.

Definition 10.1.2 (Strictly Hyperbolic System) A hyperbolic system with all distinct eigenvalues is called strictly hyperbolic.

Definition 10.1.3 (Genuinely Nonlinear Wave) Given an eigenvalue $\lambda(\mathbf{u})$ associated with the eigenvector $\mathbf{d}(\mathbf{u})$, the corresponding wave is said to be genuinely nonlinear if:

$$\nabla\lambda(\mathbf{u}) \cdot \mathbf{d}(\mathbf{u}) \neq 0 \quad \forall\, \mathbf{u},$$

where $\nabla = \partial/\partial\mathbf{u}$.

Definition 10.1.4 (Linearly Degenerate Wave) The wave is said to be exceptional or linearly degenerate if:

$$\nabla\lambda(\mathbf{u}) \cdot \mathbf{d}(\mathbf{u}) \equiv 0 \quad \forall\, \mathbf{u}. \tag{10.5}$$

We need to observe that in more than one-space dimension, λ and $\mathbf{d}$ depend also on the unit vector $\mathbf{n}$, and therefore the previous definitions need to be valid also for any $\mathbf{n}$.

10.2 Examples of Hyperbolic Systems

In this section, we verify that the Euler system for fluids, the Cattaneo heat equation, and the elasticity equation with negligible thermal conductivity are examples of hyperbolic systems.

[2] When λ becomes complex for certain field values, it can be assumed that some instability phenomena are associated with that region of the field.

10.2.1 Euler Equations

Consider the system (8.43) that governs the motion of a fluid, known as the EULER system. Choosing the state vector as

$$\mathbf{u} \equiv (\rho, v_1, v_2, v_3, S)^T,$$

the system can be written in the form (10.2) with

$$\mathbf{A}^0 \equiv \begin{pmatrix} 1&0&0&0&0\\ 0&1&0&0&0\\ 0&0&1&0&0\\ 0&0&0&1&0\\ 0&0&0&0&1 \end{pmatrix}, \quad \mathbf{A}^1 \equiv \begin{pmatrix} v_1 & \rho & 0 & 0 & 0\\ \frac{p_\rho}{\rho} & v_1 & 0 & 0 & \frac{p_S}{\rho}\\ 0 & 0 & v_1 & 0 & 0\\ 0 & 0 & 0 & v_1 & 0\\ 0 & 0 & 0 & 0 & v_1 \end{pmatrix},$$

$$\mathbf{A}^2 \equiv \begin{pmatrix} v_2 & 0 & \rho & 0 & 0\\ 0 & v_2 & 0 & 0 & 0\\ \frac{p_\rho}{\rho} & 0 & v_2 & 0 & \frac{p_S}{\rho}\\ 0 & 0 & 0 & v_2 & 0\\ 0 & 0 & 0 & 0 & v_2 \end{pmatrix}, \quad \mathbf{A}^3 \equiv \begin{pmatrix} v_3 & 0 & 0 & \rho & 0\\ 0 & v_3 & 0 & 0 & 0\\ 0 & 0 & v_3 & 0 & 0\\ \frac{p_\rho}{\rho} & 0 & 0 & v_3 & \frac{p_S}{\rho}\\ 0 & 0 & 0 & 0 & v_3 \end{pmatrix},$$

and thus, by setting $v_n = \mathbf{v} \cdot \mathbf{n}$, we have

$$\mathbf{A}_n = \mathbf{A}^i n_i \equiv \begin{pmatrix} v_n & \rho n_1 & \rho n_2 & \rho n_3 & 0\\ \frac{p_\rho}{\rho} n_1 & v_n & 0 & 0 & \frac{p_S}{\rho} n_1\\ \frac{p_\rho}{\rho} n_2 & 0 & v_n & 0 & \frac{p_S}{\rho} n_2\\ \frac{p_\rho}{\rho} n_3 & 0 & 0 & v_n & \frac{p_S}{\rho} n_3\\ 0 & 0 & 0 & 0 & v_n \end{pmatrix},$$

which has eigenvalues

$$\lambda^{(1)} = v_n - c; \quad \lambda^{(2)} = \lambda^{(3)} = \lambda^{(4)} = v_n; \quad \lambda^{(5)} = v_n + c, \tag{10.6}$$

and the corresponding eigenvectors:

$$\mathbf{d}^{(1)} \equiv (\rho, -cn_1, -cn_2, -cn_3, 0)^T, \quad \mathbf{d}^{(5)} \equiv (\rho, cn_1, cn_2, cn_3, 0)^T, \tag{10.7}$$

$$\mathbf{d}^{(2)} \equiv (-p_S, 0, 0, 0, c^2)^T, \quad \mathbf{d}^{(3)} \equiv (0, -n_3, 0, n_1, 0)^T,$$

$$\mathbf{d}^{(4)} \equiv (0, -n_2, n_1, 0, 0)^T,$$

are linearly independent if $c \neq 0$. Therefore, the EULER system with $c^2 = p_\rho > 0$ is hyperbolic in the direction of time.

Remark 9 To evaluate the eigenvalue problem (10.4) and thus verify the hyperbolicity of the system, it is not necessary to explicitly write out the matrices as done previously. In fact, by directly applying the transformation rule to the system (10.2):

$$\frac{\partial}{\partial t} \to -\lambda\delta, \qquad \frac{\partial}{\partial x_i} \to n_i\delta, \qquad \mathbf{f} \to 0 \tag{10.8}$$

and considering δ as a differential operator, we immediately have

$$(\mathbf{A}_n - \lambda\mathbf{A}^0)\delta\mathbf{u} = 0, \tag{10.9}$$

where comparing with (10.4), it follows that $\delta\mathbf{u}$ coincides with the generic right eigenvector and λ with the corresponding eigenvalue.

This method is very useful. In fact, in the previous case, without explicitly writing the matrices, we can immediately obtain from (8.43):

$$\begin{cases} (-\lambda + v_n)\delta\rho + \rho\delta v_n = 0, \\ \rho(-\lambda + v_n)\delta v_j + n_j\{p_\rho\,\delta\rho + p_S\,\delta S\} = 0, \\ (-\lambda + v_n)\delta S = 0. \end{cases} \tag{10.10}$$

By comparing (10.10) with (8.48), it can be noted that in this case the characteristic velocities coincide with the phase velocities but calculated in the general state $(\rho, \mathbf{v}, S)$.

From (10.10), by proceeding as in Sect. 8.4.3, the characteristic velocities (10.6) and the right eigenvectors (10.7) can be immediately obtained.

10.2.2 Cattaneo's Equation

In the case of a rigid conductor with equation given by Cattaneo, the first-order system consists of Eq. (9.1) with (9.3) and (9.12):

$$\begin{cases} \rho^* c_V(\vartheta) \dfrac{\partial \vartheta}{\partial t} + \dfrac{\partial q_i}{\partial x_i} = r \\[2ex] \tau(\vartheta) \dfrac{\partial q_i}{\partial t} + \chi(\vartheta) \dfrac{\partial \vartheta}{\partial x_i} = -q_i \, . \end{cases}$$

By applying the transformation rules (10.8), we obtain

$$\begin{cases} -\rho^* c_V(\vartheta) \lambda \, \delta\vartheta + \delta q_n = 0 \\[2ex] -\lambda \, \tau(\vartheta) \, \delta q_i + \chi(\vartheta) \, n_i \, \delta\vartheta = 0, \end{cases} \tag{10.11}$$

where $\delta q_n = \delta q_i \, n_i$. From (10.11), we have the following waves:

Material wave:

$$\lambda = 0; \quad \delta\vartheta = 0; \quad \delta q_n = 0.$$

The eigenvalue $\lambda = 0$ has multiplicity two. The wave does not carry temperature variations, and with respect to $\delta\mathbf{q}$, it has a transverse character, i.e., $\delta\mathbf{q}$ lies in the tangent plane orthogonal to $\mathbf{n}$.

Thermal waves: Scalar multiplying $(10.11)_2$ by $\mathbf{n}$, we have $\delta q_n = \frac{\chi}{\lambda\tau}\delta\vartheta$. Substituting this into $(10.11)_1$, we obtain

$$\delta\vartheta \text{ arbitrary}, \quad \delta\mathbf{q} = \frac{\chi}{\lambda\tau} \mathbf{n} \, \delta\vartheta, \quad \lambda = \pm\sqrt{\frac{\chi(\vartheta)}{\rho^* \, c_V(\vartheta) \, \tau(\vartheta)}} \, .$$

Choosing the field $\mathbf{u} \equiv (\vartheta, \mathbf{q})^T$ and recalling from (10.9) that $\mathbf{d} \propto \delta\mathbf{u}$, we can summarize that the following eigenvalues and eigenvectors exist:

$$\lambda^{(1)} = -\sqrt{\frac{\chi}{\rho^* \, c_V \, \tau}}; \qquad \lambda^{(2)} = \lambda^{(3)} = 0; \qquad \lambda^{(4)} = \sqrt{\frac{\chi}{\rho^* \, c_V \, \tau}};$$

$$\mathbf{d}^{(1)} \equiv \left(1, \frac{\chi}{\lambda_1 \tau} \mathbf{n}\right)^T; \quad \mathbf{d}^{(2)} \equiv (0, \mathbf{w}_1)^T;$$

$$\mathbf{d}^{(3)} \equiv (0, \mathbf{w}_2)^T; \quad \mathbf{d}^{(4)} \equiv \left(1, \frac{\chi}{\lambda_4 \tau} \mathbf{n}\right)^T,$$

where $\mathbf{w}_1$ and $\mathbf{w}_2$ satisfy $\mathbf{w}_1 \cdot \mathbf{n} = \mathbf{w}_2 \cdot \mathbf{n} = \mathbf{w}_1 \cdot \mathbf{w}_2 = 0$. As we can see, the system is hyperbolic as long as the function $\tau(\vartheta)$ is positive. In the case of $\tau \to 0$, we have $\lambda \to \pm\infty$, which corresponds to the parabolic case with the paradox of infinite propagation velocity.

The Cattaneo model, although able to eliminate the heat paradox, remains only an approximation. For a rigorous and complete investigation, a much more sophisticated analysis is required, which falls under the field of *Rational Extended Thermodynamics* (see Müller and Ruggeri [19], Ruggeri and Sugiyama [20]).

10.2.3 Thermoelasticity Equations

Let us consider the system of thermoelasticity (7.14) with zero thermal conductivity ($\chi = 0$) and thus no heat flux $\mathbf{Q} = 0$, and as consequence, $\Sigma^* = 0$, see (7.31).

In this case, it is convenient to replace the energy equation $(7.14)_3$ with the entropy equation (6.8). Therefore, we consider the following system:

$$\begin{cases} \rho^* \dfrac{\partial v_i}{\partial t} - \dfrac{\partial T_{iC}}{\partial X_C} = \rho^* b_i \\[2ex] \dfrac{\partial F_{iA}}{\partial t} - \dfrac{\partial v_i}{\partial X_A} = 0 \\[2ex] \rho^* \dfrac{\partial S}{\partial t} = \dfrac{r^*}{\vartheta} \end{cases} . \tag{10.12}$$

Since the system (10.12) is written in Lagrangian variables, the transformation rule for operators (10.8) becomes

$$\frac{\partial}{\partial t} \to -\lambda\delta, \qquad \frac{\partial}{\partial X_A} \to N_A \delta, \qquad \mathbf{f} \to 0. \tag{10.13}$$

Applying (10.8) to (10.12), we obtain

$$\begin{cases} \rho^* \lambda \delta v_i + \delta T_{iC} N_C = 0 \\[2ex] \lambda \delta F_{iA} + N_A \delta v_i = 0 \\[2ex] \lambda \delta S = 0 \end{cases} . \tag{10.14}$$

From (10.14), we can deduce the presence of two types of *waves*:

Contact wave: $\lambda = 0$. This wave has the following properties:

$$\delta \mathbf{v} = 0; \quad \delta \mathbf{T} \cdot \mathbf{N} = 0; \quad \delta S \text{ arbitrary}.$$

Sound waves: $\lambda \neq 0$. From $(10.14)_3$, we have $\delta S = 0$. Eliminating δv_i from $(10.14)_1$ and substituting it into $(10.14)_2$ (taking into account that T_{iA} is a function of F_{jB}), we obtain

$$\left(\rho^* \lambda^2 \delta_{ij} \delta_{AB} - \frac{\partial T_{iC}}{\partial F_{jB}} N_C N_A\right) \delta F_{jB} = 0.$$

Using the notation (7.42), we have

$$\left(\lambda^2 \delta_{ij} \delta_{AB} - \frac{\partial^2 e}{\partial F_{iC} \partial F_{jB}} N_C N_A\right) \delta F_{jB} = 0. \tag{10.15}$$

Applying the transformation rule (10.13) to the successive derivatives and considering (2.3), we can write

$$\delta F_{jB} = N_B \delta^2 x_j.$$

Therefore, (10.15) becomes

$$(K_{ij} - \lambda^2 \, \delta_{ij}) \, \delta^2 \, x_j = 0 \quad \Leftrightarrow \quad \left(\mathbf{K} - \lambda^2 \mathbf{I}\right) \delta^2 \mathbf{x} = 0, \tag{10.16}$$

where $\mathbf{K}$ is the operator with components:

$$K_{ij} = \frac{\partial^2 e}{\partial F_{iC} \partial F_{jB}} N_C N_B.$$

Thus, the differential system is hyperbolic if the operator $\mathbf{K}$ is positive-definite, meaning that e is a convex function of $\mathbf{F}$. However, this condition of convexity holds only near an equilibrium state, i.e., for small deformations. In fact TRUESDELL proved that the global convexity is in contrast with the principle of material indifference [2]. In the linearized case from (7.62), using the transformation rule (10.8), we have

$$(\rho \lambda^2 - \mu) \delta^2 u_i - (\nu + \mu) n_i n_j \delta^2 u_j = 0.$$

This equation can be written in vector form as

$$(\rho \lambda^2 - \mu) \delta^2 \mathbf{u} - (\nu + \mu) \mathbf{n} \, \delta^2 u_n = 0. \tag{10.17}$$

Here, $\delta^2 u_n = \delta^2 \mathbf{u} \cdot \mathbf{n}$. From (10.17), we have

$$\lambda = \pm\sqrt{\frac{\mu}{\rho}}; \qquad \delta^2 u_n = 0 \quad \text{(transverse waves)}$$

or

$$\lambda = \pm\sqrt{\frac{\nu + 2\mu}{\rho}}; \qquad \delta^2 \mathbf{u} \parallel \mathbf{n} \quad \text{(longitudinal waves).}$$

For hyperbolicity, it is required that

$$\mu > 0 \quad \text{and} \quad \nu + 2\mu > 0,$$

which is consistent with the previous results obtained for plane wave solutions (see Sect. 7.63).

10.3 Wave Equation and Method of Characteristics

Let us now consider a linear hyperbolic system of the form (10.2) in one spatial dimension without production terms:

$$\mathbf{u}_t + \mathbf{A}\mathbf{u}_x = 0, \tag{10.18}$$

where indices indicate differentiation with respect to the arguments. An example of (10.18) is the wave equation (7.67), which can be rewritten as

$$U_{tt} - c^2 U_{xx} = 0. \tag{10.19}$$

To transform this into a first-order system of the form (10.18), we set

$$U_t = v, \quad U_x = w. \tag{10.20}$$

Equation (10.19), along with the compatibility condition of SCHWARTZ for (10.20), yields

$$\begin{cases} v_t - c^2 w_x = 0 \\ w_t - v_x = 0 \end{cases}. \tag{10.21}$$

Taking

$$\mathbf{u} = (v, w)^T, \tag{10.22}$$

the system (10.21) fits into (10.18) with

$$\mathbf{A} = \begin{pmatrix} 0 & -c^2 \\ -1 & 0 \end{pmatrix}. \tag{10.23}$$

The eigenvalue problem in this case is trivial:

$$\lambda^{(1)} = -c; \qquad \lambda^{(2)} = c; \tag{10.24}$$

and

$$\mathbf{d}^{(1)} \equiv \begin{pmatrix} c \\ 1 \end{pmatrix}; \qquad \mathbf{d}^{(2)} \equiv \begin{pmatrix} -c \\ 1 \end{pmatrix}. \tag{10.25}$$

Let $\mathbf{l}$ denote the general left eigenvector of $\mathbf{A}$, that is, the transpose of the right eigenvector of $\mathbf{A}^T$:

$$\mathbf{l}\,\mathbf{A} = \lambda \mathbf{l} \tag{10.26}$$

($\mathbf{l}$ is a row vector $(1 \times N)$ to be multiplied to the left of the operator $\mathbf{A}$). In the present case, we have

$$\mathbf{l}^{(1)} \equiv (1, c)\,; \qquad \mathbf{l}^{(2)} \equiv (1, -c). \tag{10.27}$$

Therefore, the system (10.21) is strictly hyperbolic since all eigenvalues are distinct. It is also worth noting that since the system (10.21) is linear, the eigenvalues and eigenvectors found are all constants.

Consider Eq. (10.19) in an unbounded domain. Take the Cauchy data (initial data) as

$$U(x, 0) = \varphi(x), \quad U_t(x, 0) = \psi(x). \tag{10.28}$$

Referring to the first-order system (10.21) (taking into account the conditions (10.20)), we have

$$v(x, 0) = \psi(x), \quad w(x, 0) = \varphi'(x); \quad \text{where } \varphi'(x) = \frac{d\varphi}{dx}. \tag{10.29}$$

Hence, Eq. (10.19) with the conditions (10.28) is equivalent to studying the system (10.21) with the initial conditions (10.29).

Now we multiply (10.18) by a generic left eigenvector of $\mathbf{A}$ (as in (10.26)):

$$\mathbf{l}\{\mathbf{u}_t + \mathbf{A}\mathbf{u}_x\} = 0 \quad \Leftrightarrow \quad \mathbf{l}\{\mathbf{u}_t + \lambda \mathbf{u}_x\} = 0, \tag{10.30}$$

and we define the *characteristic curve* as the line in space-time satisfying:

$$\frac{dx}{dt} = \lambda. \tag{10.31}$$

In the present case, there are two characteristic curves (note that $\lambda = \pm c$), and they are straight lines since the values of λ are constant:

$$C^1 \Rightarrow \frac{dx}{dt} = \lambda^{(1)} \Rightarrow x = -ct + x_1 \tag{10.32}$$

$$C^2 \Rightarrow \frac{dx}{dt} = \lambda^{(2)} \Rightarrow x = ct + x_2. \tag{10.33}$$

The two lines have slopes of $-c$ and $+c$. Now, let us calculate the directional derivative:

$$\frac{d}{dt} = \partial_t + \lambda\, \partial_x \tag{10.34}$$

along one of these characteristic curves. By doing so, we can immediately see that (10.30) can be rewritten:

$$\mathbf{l} \cdot \frac{d\mathbf{u}}{dt} = 0 \tag{10.35}$$

and for each characteristic curve:

$$\mathbf{l}^{(1)} \cdot \frac{d^{(1)}\mathbf{u}}{dt} = 0 \quad (\text{on } C^1), \qquad \mathbf{l}^{(2)} \cdot \frac{d^{(2)}\mathbf{u}}{dt} = 0 \quad (\text{on } C^2). \tag{10.36}$$

Since $\mathbf{l}^{(1)}$ and $\mathbf{l}^{(2)}$ are constants, it follows that

$$\frac{d}{dt}(\mathbf{l}^{(1)} \cdot \mathbf{u}) = 0 \quad (\text{on } C^1),$$
$$\frac{d}{dt}(\mathbf{l}^{(2)} \cdot \mathbf{u}) = 0 \quad (\text{on } C^2). \tag{10.37}$$

This means that for any given point $P \equiv (x, t)$ in space-time (see Fig. 10.1), $\mathbf{l}^{(1)} \cdot \mathbf{u}$ is constant along C^1 and $\mathbf{l}^{(2)} \cdot \mathbf{u}$ is constant along C^2; therefore

$$\mathbf{l}^{(1)} \cdot \mathbf{u}(x, t) = \mathbf{l}^{(1)} \cdot \mathbf{u}(x_1, 0) \tag{10.38}$$

$$\mathbf{l}^{(2)} \cdot \mathbf{u}(x, t) = \mathbf{l}^{(2)} \cdot \mathbf{u}(x_2, 0). \tag{10.39}$$

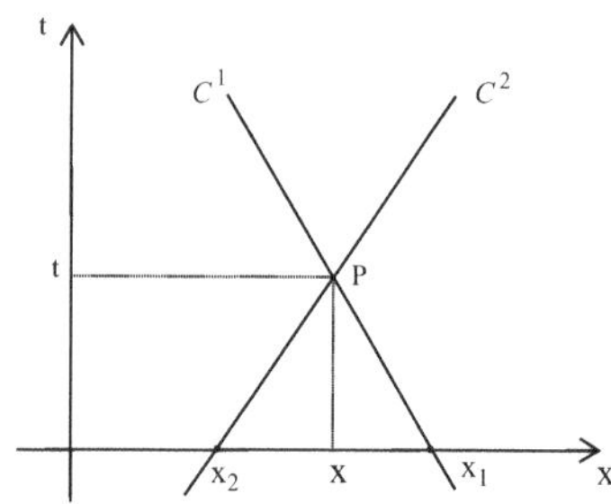

Fig. 10.1 Characteristic lines for the wave equation

Let $\mathbf{u}_0(x)$ be the initial data for $\mathbf{u}$:

$$\mathbf{u}(x,0)=\mathbf{u}_0(x).$$

Taking into account equations (10.32) and (10.33), Eqs. (10.38) and (10.39) become

$$\begin{cases} \mathbf{l}^{(1)}\cdot\mathbf{u}(x,t)=\mathbf{l}^{(1)}\cdot\mathbf{u}_0(x+ct) \\ \\ \mathbf{l}^{(2)}\cdot\mathbf{u}(x,t)=\mathbf{l}^{(2)}\cdot\mathbf{u}_0(x-ct). \end{cases} \tag{10.40}$$

Equations (10.40) are two scalar equations in the components v and w. By solving this algebraic system, we obtain v and w in terms of the initial data v_0 and w_0. This solves the proposed problem. By explicitly expressing equations (10.40) considering (10.27), (10.22), and (10.29), we obtain the algebraic system:

$$\begin{cases} v(x,t)+c\,w(x,t)=\psi(x+ct)+c\varphi'(x+ct) \\ \\ v(x,t)-c\,w(x,t)=\psi(x-ct)-c\varphi'(x-ct), \end{cases}$$

and therefore the solution

$$v(x,t)=\frac{1}{2}\left\{\psi(x+ct)+\psi(x-ct)+c\left[\varphi'(x+ct)-\varphi'(x-ct)\right]\right\} \tag{10.41}$$

$$w(x,t)=\frac{1}{2c}\left\{\psi(x+ct)-\psi(x-ct)+c\left[\varphi'(x+ct)+\varphi'(x-ct)\right]\right\}. \tag{10.42}$$

In terms of the field U, from $(10.20)_2$, we have

$$U(x,t)=\frac{1}{2c}\left\{\Gamma(x+ct)-\Gamma(x-ct)+c\left[\varphi(x+ct)+\varphi(x-ct)\right]\right\} \tag{10.43}$$

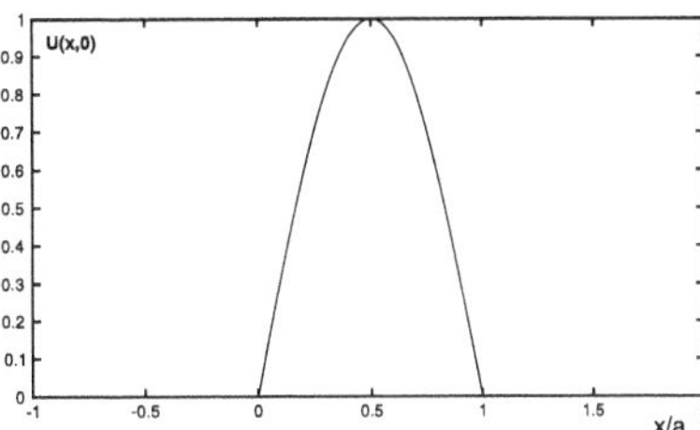

Fig. 10.2 Initial data for pinched string

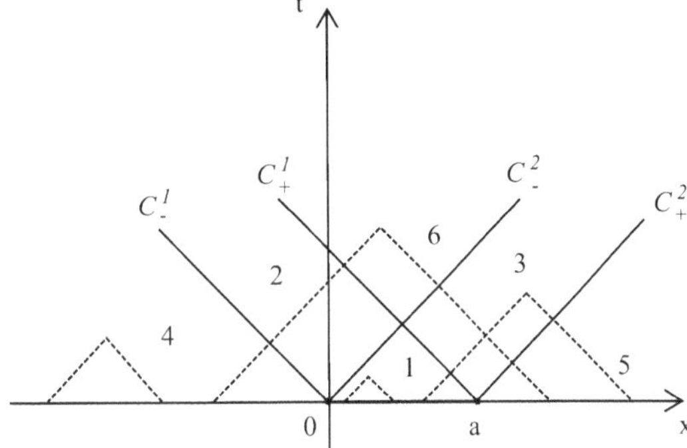

Fig. 10.3 Space-time regions for the pinched string

with $\Gamma(\xi)$ a primitive of $\psi(\xi)$

$$\Gamma(\xi) = \int \psi(\xi)\, d\xi.$$

Equation (10.43) represents the classical solution of the one-dimensional wave equation. Essentially, the method illustrated boils down to finding the *Riemann invariants*, which are quantities that remain constant along the characteristics. Now let us consider a particular case: the *pinched string*. Suppose the initial data is given by

$$\begin{cases} \psi(x) = 0, \\ \varphi(x) \text{ a continuous function that is zero} \\ \text{outside the interval } x \in [0, a] \end{cases}. \tag{10.44}$$

The condition $\psi(x) = 0$ means that initially there is no velocity, while the condition on $\varphi(x)$ means that at the initial time, the signal is localized within a certain interval $(0 < x < a)$ as shown in Fig. 10.2.

To study the evolution from this initial data, we draw the characteristics from the points $(0, 0)$ and $(a, 0)$ in space-time, dividing the semi-plane $(-\infty < x < \infty, t \geq 0)$ into six domains (see Fig. 10.3). It can be observed that for any point in regions 5, 6, 4, when we construct the characteristic triangle (i.e., the triangle formed by drawing lines parallel to the characteristics C^1_+, C^1_-, C^2_+, C^2_- from the given point until they intersect the x-axis), it does not intersect the segment $[0, a]$ where the initial conditions (10.44) are nonzero. In these domains, the solution is therefore zero (in accordance with $\varphi(x) = 0$ for $x \le 0,\ x \ge a$ given in (10.44)). Physically, this means that for large t (region 6), the signal has already passed, while for large x (regions 4 and 5), no signal has arrived yet (the initial perturbation has not reached these regions). The situation is different in domains 2 and 3: The characteristic triangle for points in these regions intersects (partially) the segment $[0, a]$. In domain 3, the solution is the *right-going wave* (see (10.43)):

$$U(x, t) = \frac{1}{2}\varphi(x - ct). \tag{10.45}$$

In domain 2, the solution is the *left-going wave* (see (10.43)):

$$U(x, t) = \frac{1}{2}\varphi(x + ct). \tag{10.46}$$

The remaining domain is 1: The characteristic triangle for points in this region always intersects the segment $[0, a]$. The solution (see (10.43)) is a linear combination of the two waves, i.e., the sum of the left-going and right-going waves:

$$U(x, t) = \frac{1}{2}\left\{\varphi(x + ct) + \varphi(x - ct)\right\}. \tag{10.47}$$

Physically, we can interpret the previous scenario as follows: *For sufficiently small times (domain* 1*), the perturbation is a combination of the two waves. As time increases, the perturbation splits into two waves propagating in opposite directions (domains* 2 *and* 3*), as shown in Fig. 10.4. For very large values of x (domains* 4 *and* 5*), the perturbation cannot be seen yet because it has not arrived, while in domain* 6*, corresponding to long times, the signal has already passed.*

An example of evolution in time is reported in Fig. 10.4 with initial data given in Fig. 10.2.

10.3.1 Homogeneous Linear Systems

Returning to the method of characteristics illustrated in the previous pages, let us see how to generalize it when the number of linear equations in (10.18) is N ($\mathbf{u} \in \mathbb{R}^N$). Multiplying (10.18) by a generic left eigenvector $\mathbf{l}^{(i)}$ of $\mathbf{A}$, we obtain from (10.26)

$$\mathbf{l}^{(i)} \cdot \left\{\mathbf{u}_t + \lambda^{(i)}\mathbf{u}_x\right\} = 0, \quad i = 1, 2, \ldots, N. \tag{10.48}$$

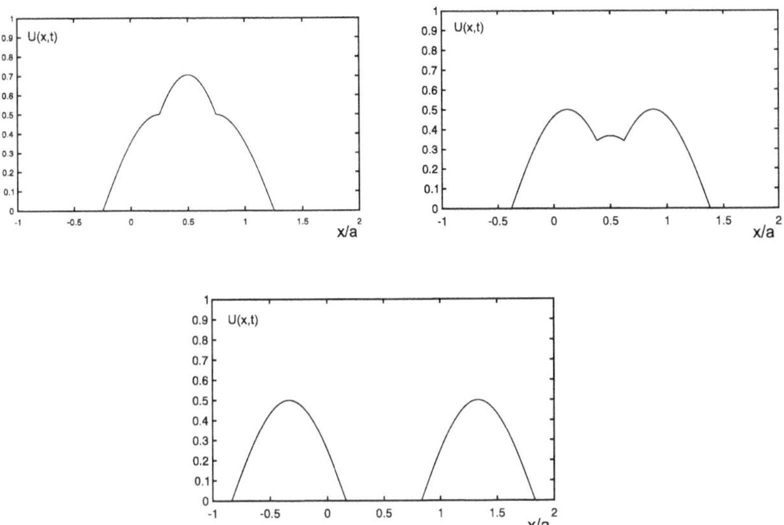

Fig. 10.4 Evolution at time t: Formation of two waves

Taking into account that $dx/dt = \lambda^{(i)}$ and thus $x = \lambda^{(i)}t + x_i$, we have

$$\frac{d}{dt}(\mathbf{l}^{(i)} \cdot \mathbf{u}) = 0 \quad \Longrightarrow \quad \mathbf{l}^{(i)} \cdot \mathbf{u}(x,t) = \text{constant} \quad (\text{on } C^{(i)}). \tag{10.49}$$

Since the right (and left) eigenvectors form a basis of the vector space, the vector $\mathbf{u}(x, t)$ can always be written as

$$\mathbf{u}(x,t) = \sum_{j=1}^{N} \Pi^j(x,t)\mathbf{d}^{(j)}, \tag{10.50}$$

where the unknowns are the $\Pi^j(x, t)$. Substituting (10.50) into the second equation of (10.49), and considering that the right and left eigenvectors can be orthonormalized ($\mathbf{l}^{(i)} \cdot \mathbf{d}^{(j)} = \delta^{ij}$) and that similarly to (10.50), we also have

$$\mathbf{u}_0(x) = \sum_{j=1}^{N} \Pi_0^j(x)\mathbf{d}^{(j)},$$

and we obtain

$$\Pi^i(x,t) = \Pi_0^i(x_i) = \Pi_0^i(x - \lambda^i t). \tag{10.51}$$

By substituting (10.51) into (10.49), we have

$$\mathbf{u}(x,t) = \sum_{j=1}^{N} \Pi_0^j(x - \lambda^j t)\mathbf{d}^{(j)}. \tag{10.52}$$

This solution solves the problem of determining **u** using the initial conditions. Equation (10.52) tells us that **u** *is a superposition of N waves traveling with characteristic velocities*.

10.4 A Nonlinear Example: The Burgers' Equation

The nonlinear case is fundamentally different from the linear case, especially because, in general, there are no regular solutions for all times. As an example, let us consider the scalar nonlinear equation known as the BURGERS' equation:

$$u_t + uu_x = 0. \tag{10.53}$$

It is evident that in this case the only eigenvalue is $\lambda = u$. Similarly to what we did before, we can define the characteristic curve, but only in a formal way:

$$\frac{dx}{dt} = u\,(x,t)\,, \tag{10.54}$$

where u is the unknown function of our problem. However, from Eq. (10.53), using Eq. (10.34), we have

$$\frac{du}{dt} = u_t + \lambda u_x = u_t + uu_x = 0. \tag{10.55}$$

Therefore, on the characteristic curve C, u remains constant and equal to its initial value at the point x_0 where the characteristic curve intersects the x-axis:

$$u(x,t) = u_0(x_0). \tag{10.56}$$

By inserting (10.56) into (10.54), we find that the characteristic curve is necessarily a straight line given by

$$x = x_0 + u_0(x_0)\,t \tag{10.57}$$

(see Fig. 10.5a). If the Cauchy data (10.56) is not constant, considering two points x_1 and x_2 at time $t = 0$, where $u_0(x_1) \neq u_0(x_2)$, the two characteristic curves passing through x_1 and x_2 will not be parallel and will intersect at a point P in space-time (see Fig. 10.5b). The solution at P, belonging to the characteristic curve passing

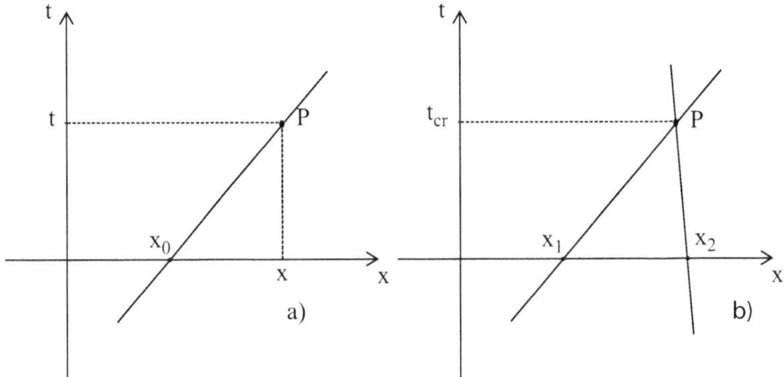

Fig. 10.5 Characteristic curve and critical time in the Burgers' equation

through x_1, is given by $u(x, t) = u_0(x_1)$, and belonging to the characteristic curve passing through x_2, it is $u(x, t) = u_0(x_2)$. In point P, the uniqueness of the solution is lost. *The smallest time in the future where the uniqueness of the solution is lost is called the critical time* (t_{cr}). Before t_{cr}, there exists a regular solution of (10.53) (*classical solution*), but after t_{cr}, since the solution is no longer differentiable, a *shock wave* is typically formed, which is a particular type of *weak solution* as we will see in the next chapter.

The determination of the critical time is straightforward for the Burgers' equation. In fact, (10.56) and (10.57) represent the solutions in parametric form: For a given time t, the dependence of u on x is through the parameter x_0. Equation (10.57) provides x_0 as a function of x, provided that $dx/dx_0 \neq 0$. Therefore, invertibility is lost when $dx/dx_0 = 0$, and we have that each point x_0 of the initial profile $u(x_0)$ will have a critical time (in the future or in the past) given by

$$t_c(x_0) = -\frac{1}{u'(x_0)}.$$

The critical time t_{cr} is the smallest positive value of $t_c(x_0)$:

$$t_{cr} = \inf_{x_0}\{t_c(x_0) > 0\}. \tag{10.58}$$

For example, if we take as initial data the one represented in Fig. 10.6:

$$u_0(x) = \begin{cases} \sin(\pi x/a) & \text{for} \quad x \in [0, a], \\ 0 & \text{for} \quad x \notin [0, a]. \end{cases}$$

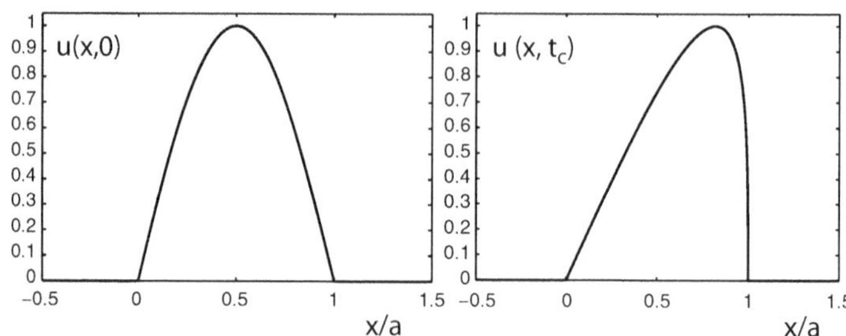

Fig. 10.6 Critical time in the Burgers' equation

Equations (10.56) and (10.57) become

$$\begin{cases} u(x,t) = \sin(\pi x_0/a); \quad x \notin [0,a] \\ \\ x = x_0 + t\sin(\pi x_0/a) \end{cases},$$

and the critical time (10.58) is given by

$$t_c = \frac{a}{\pi}\,.$$

From Fig. 10.6, it can be observed how the profile deforms as time varies. This is due to the fact that the characteristic velocity $\lambda = u$ and, consequently, the highest points of the profile travel faster. At the critical time, the derivative becomes unbounded at some point (in this case, at $x_c = x_0 = a$).

Chapter 11
Weak Solutions and Shock Waves

Abstract Among the phenomena of nonlinear propagation, *shock waves* play a particular role. Through a moving front that separates space into two half-spaces, the field variables undergo a discontinuity. To study shock waves, it is necessary to extend the concept of solutions to include the case of discontinuous functions. For this reason, the concept of *weak solution* is introduced in this chapter. A brief overview of shock wave theory and the issue of nonuniqueness for weak solutions, particularly in the context of the RIEMANN problem, will be discussed. As an example of this problem, we will examine shock waves in Euler fluid dynamics and the fluidodynamic model of *car traffic*.

11.1 Shock Waves and Weak Solutions

Consider a system of balance laws given by

$$\partial_\alpha \mathbf{F}^\alpha(\mathbf{u}) = \mathbf{f}(\mathbf{u}). \tag{11.1}$$

If there exists a smooth surface Γ with a normal unit vector $\mathbf{n}$ that moves with a normal velocity s, separating two half-spaces where two continuous solutions of (11.1), $\mathbf{u}_0$ and $\mathbf{u}_1$, exist, but their limits on the surface do not coincide, then the pair $(\mathbf{u}_0, \mathbf{u}_1)$ represents a *shock wave*, and the shock wave is identified by its wavefront (see Fig. 11.1).

We denote the jump with square brackets:

$$[\mathbf{u}] = \mathbf{u}_1 \big|_{\varphi^-} - \mathbf{u}_0 \big|_{\varphi^+} \quad (\text{on } \Gamma).$$

First, let us recall the notion of weak solution. Suppose that a domain C in space-time is given (Fig. 11.2).

© The Author(s), under exclusive license to Springer Nature Switzerland AG 2024

T. Ruggeri, *Introduction to the Thermomechanics of Continua and Hyperbolic Systems*, La Matematica per il 3+2 167,
https://doi.org/10.1007/978-3-031-69951-1_11

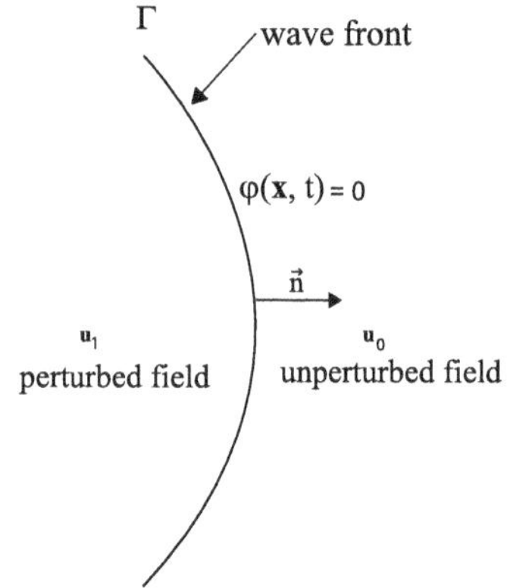

Fig. 11.1 Shock wave

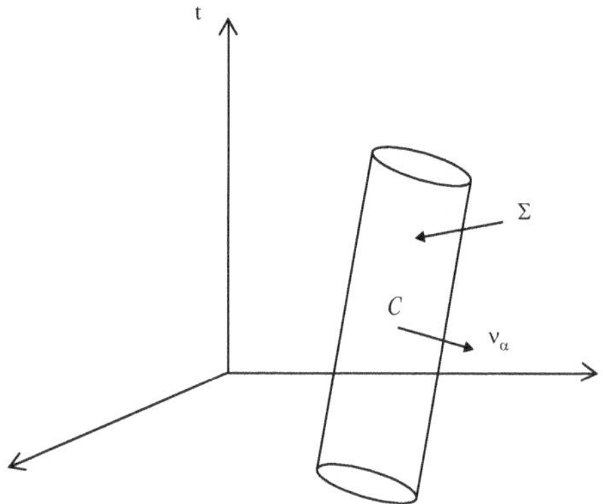

Fig. 11.2 Weak solutions

We multiply (11.1) by a test function ϕ with support in C (a C^∞ function that is zero on the boundary of C) and integrate over the domain C. We have

$$\int_C \phi \left(\partial_\alpha \mathbf{F}^\alpha - \mathbf{f}\right) dC = 0, \tag{11.2}$$

which can also be written as

$$\int_C \partial_\alpha \left(\phi \mathbf{F}^\alpha\right) dC - \int_C \left(\mathbf{F}^\alpha \partial_\alpha \phi + \mathbf{f}\phi\right) dC = 0. \tag{11.3}$$

Now, let us apply the GAUSS–GREEN theorem to the first integral of (11.3). Denoting by Σ the surface of C and by ν_α the normal vector (in space-time) to that surface, we obtain from (11.3)

$$\int_\Sigma \nu_\alpha \left(\phi \mathbf{F}^\alpha\right) d\Sigma - \int_C (\mathbf{F}^\alpha \partial_\alpha \phi + \mathbf{f}\phi) dC = 0. \tag{11.4}$$

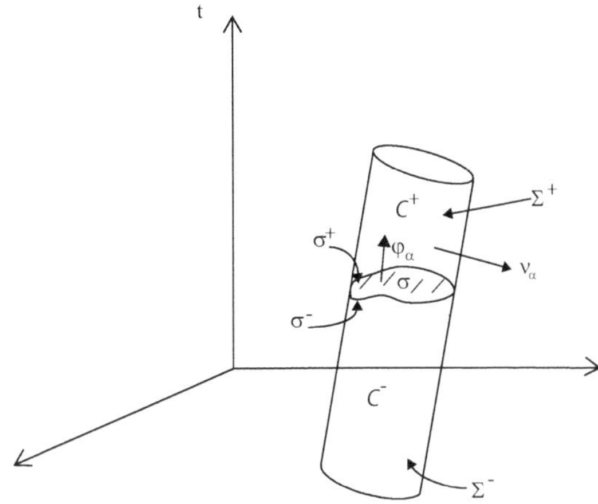

Fig. 11.3 Shock waves as weak solutions

Since ϕ has compact support, the first integral is zero (assuming, for simplicity, zero initial data). Thus, we have

$$\int_C \mathbf{F}^\alpha \partial_\alpha \phi dC + \int_C \mathbf{f}\phi dC = 0. \tag{11.5}$$

We say that a solution of (11.1) *is weak if it satisfies* (11.5) *for every test function ϕ with compact support.*

From the procedure described, we can deduce that *every classical solution is weak, but not vice versa*. Now, let us consider the case of a shock wave and prove that if the pair $\mathbf{u}_0$, $\mathbf{u}_1$ is a weak solution, then certain compatibility equations, known as RANKINE–HUGONIOT equations (R-H equations), must be satisfied at the shock front.

For this purpose, we consider a surface σ with normal φ_α that moves within the domain C and divides it into two subdomains, C^+ and C^-, with boundaries Σ^+ and Σ^-, respectively (Fig. 11.3).

We now consider (11.4) in each of the subdomains C^+ and C^- and apply the GAUSS–GREEN theorem. We obtain the following equations in C^+ and C^-, respectively:

$$\int_{\Sigma^+ \cup \sigma^+} \varphi_\alpha^+ \phi \mathbf{F}_+^\alpha d\Sigma^* - \int_{C^+} (\mathbf{F}_+^\alpha \partial_\alpha \phi + \mathbf{f}_+ \phi) dC^+ = 0, \tag{11.6}$$

$$\int_{\Sigma^- \cup \sigma^-} \varphi_\alpha^- \phi \mathbf{F}_-^\alpha d\Sigma^* - \int_{C^-} (\mathbf{F}_-^\alpha \partial_\alpha \phi + \mathbf{f}_- \phi) dC^- = 0. \tag{11.7}$$

Since ϕ has compact support in C^+, only the integral extended to σ^+ survives in (11.6). Similarly, in (11.7), only the integral extended to σ^- survives. Thus, we can rewrite (11.6) and (11.7) as follows:

$$\int_{\sigma^+} \varphi_\alpha^+ \phi \mathbf{F}_+^\alpha d\sigma^+ - \int_{C^+} (\mathbf{F}_+^\alpha \partial_\alpha \phi + \mathbf{f}_+ \phi) dC^+ = 0, \tag{11.8}$$

$$\int_{\sigma^-} \varphi_\alpha^- \phi \mathbf{F}_-^\alpha d\sigma^- - \int_{C^-} (\mathbf{F}_-^\alpha \partial_\alpha \phi + \mathbf{f}_- \phi) dC^- = 0. \tag{11.9}$$

By summing (11.8) and (11.9), we obtain

$$\int_{\sigma^+} \varphi_\alpha^+ \phi \mathbf{F}_+^\alpha d\sigma^+ + \int_{\sigma^-} \varphi_\alpha^- \phi \mathbf{F}_-^\alpha d\sigma^- - \int_C (\mathbf{F}^\alpha \partial_\alpha \phi + \mathbf{f}\phi) dC = 0. \tag{11.10}$$

The last integral is zero because we assume that the sought solution is weak, and thus (11.5) is satisfied. Now, we observe that $\varphi_\alpha^+ = \varphi_\alpha$ and $\varphi_\alpha^- = -\varphi_\alpha$, so we can rewrite (11.10) as

$$\int_\sigma \varphi_\alpha (\mathbf{F}_+^\alpha - \mathbf{F}_-^\alpha) \phi \, d\sigma = 0. \tag{11.11}$$

Since this integral must vanish for any surface σ, we have

$$(\mathbf{F}_+^\alpha - \mathbf{F}_-^\alpha) \varphi_\alpha = 0, \tag{11.12}$$

which can also be written as

$$[\mathbf{F}^\alpha] \varphi_\alpha = 0. \tag{11.13}$$

The (11.13) implies that the normal component in space-time of $\mathbf{F}^\alpha$ is continuous, and therefore we can conclude that:

For a shock wave to be a weak solution, the normal component of $\mathbf{F}^\alpha$ *must be continuous across the shock front* Γ.

Eqs. (11.13) are called *Rankine–Hugoniot equations* (R-H). Now, let us separate the temporal part from the spatial part of the normal vector φ_α:

$$\varphi_0 = -s, \qquad \varphi_i = n_i, \tag{11.14}$$

where s is the normal velocity of the shock wave and $\mathbf{n} \equiv (n_i)$ is the normal unit vector to the front. Assuming $\mathbf{u} \equiv \mathbf{F}^0$, we can write (11.12) as

$$-s\,[\mathbf{u}] + \left[\mathbf{F}^i\right] n_i = 0 \tag{11.15}$$

or

$$-s\mathbf{u}_1 + \mathbf{F}^i(\mathbf{u}_1) n_i = -s\mathbf{u}_0 + \mathbf{F}^i(\mathbf{u}_0) n_i, \tag{11.16}$$

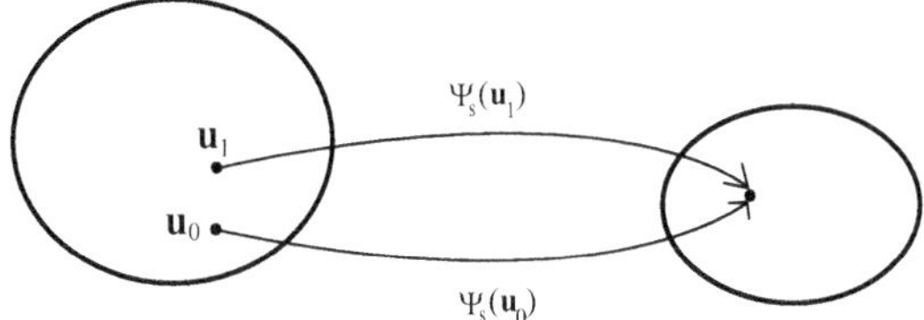

Fig. 11.4 Non-invertibility for the existence of shock waves

where, for brevity, we have denoted by $\mathbf{u}_1$ and $\mathbf{u}_0$ the values of the left and right limits of $\mathbf{u}$ on the surface.

Writing the Rankine–Hugoniot equations is simple: We formally use the *operator rule* in the field equations:

$$\partial_t \to -s[\cdot], \quad \partial_i \to n_i[\cdot], \quad \mathbf{f} \to 0. \tag{11.17}$$

Now, let us return to (11.16) and consider the application (dependent on the parameter s):

$$\boldsymbol{\Psi}_s(\mathbf{u}) = -s\mathbf{u} + \mathbf{F}^i(\mathbf{u})n_i. \tag{11.18}$$

Equation (11.16) becomes[1]

$$\boldsymbol{\Psi}_s(\mathbf{u}_1) = \boldsymbol{\Psi}_s(\mathbf{u}_0). \tag{11.19}$$

If the function $\boldsymbol{\Psi}_s(\mathbf{u})$ is invertible, (11.19) would have the unique solution $\mathbf{u}_1 = \mathbf{u}_0$, which corresponds to no shock. Therefore, to have a nontrivial shock, we must require the lack of local invertibility for $\boldsymbol{\Psi}_s(\mathbf{u})$ (see Fig. 11.4). Let us calculate the Jacobian matrix of the transformation (11.18). We have

$$\frac{\partial \boldsymbol{\Psi}_s}{\partial \mathbf{u}} = -s\mathbf{I} + \mathbf{A}^i n_i, \tag{11.20}$$

where we have indicated $\mathbf{A}^i = \partial \mathbf{F}^i/\partial \mathbf{u}$. In order to have a nontrivial shock, there must exist values of the field $\mathbf{u}_0$ and the parameter s such that the determinant of (11.20) is zero, i.e., s must coincide with an eigenvalue, since (see (10.4))

$$\det(\mathbf{A}^i n_i - \lambda \mathbf{I}) = 0. \tag{11.21}$$

Shock waves can be seen as *branches of bifurcation* from the trivial solution $\mathbf{u}_1 = \mathbf{u}_0$ as represented in Fig. 11.5. Shocks that bifurcate for $s = \lambda_0^{(k)}$ are called *k-shocks* and are denoted by S_k ($k = 1, 2, \ldots, N$). Fixing a k, the Rankine–Hugoniot equations (assuming $\mathbf{u}_0$ and $\mathbf{n}$ known) are N scalar equations for the $N+1$ unknowns

[1] Note that there always exists for any s the trivial solution (no shock) $\mathbf{u}_1 = \mathbf{u}_0 \Leftrightarrow [\mathbf{u}] = 0$.

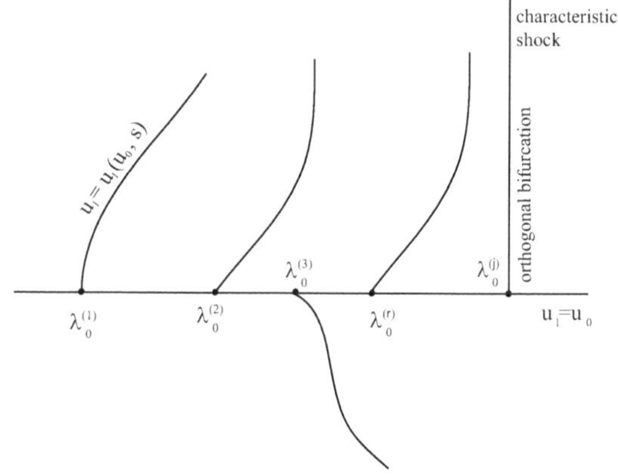

Fig. 11.5 Shocks as bifurcation of the null shock

$\mathbf{u}_1$ and s. Therefore, the solutions are (in general) a one-parameter family; for example, choosing s as the parameter, we have

$$\mathbf{u}_1 \equiv \mathbf{u}_1(\mathbf{u}_0, s, \mathbf{n}). \tag{11.22}$$

11.2 Shock Waves in Eulerian Fluid

As an example of Rankine–Hugoniot equations, let us consider the Euler fluid. By using the formal substitution (11.17), we can easily write the R-H equations for the system (8.5). We have

$$-s\,[\rho] + [\rho v_n] = 0 \tag{11.23}$$

$$-s\,[\rho\,\mathbf{v}] + [\rho\, v_n\, \mathbf{v} + p\,\mathbf{n}] = 0 \tag{11.24}$$

$$-s\left[\rho\frac{v^2}{2} + \rho e\right] + \left[\left(\rho\frac{v^2}{2} + \rho e + p\right) v_n\right] = 0, \tag{11.25}$$

where $v_n = \mathbf{v}\cdot\mathbf{n}$.

Explicitly, Eqs. (11.23)–(11.25) can be written as (from now on, we omit the subscript 1 for the perturbed field)

$$(-s + v_n)\,\rho = (-s + v_{0n})\,\rho_0 \tag{11.26}$$

$$(-s + v_n)\,\rho\,\mathbf{v} + p\,\mathbf{n} = (-s + v_{0n})\,\rho_0\,\mathbf{v}_0 + p_0\mathbf{n} \tag{11.27}$$

$$(-s + v_n)\left(\rho\frac{v^2}{2} + \rho\, e\right) + p\, v_n = (-s + v_{0n})\left(\rho_0\frac{v_0^2}{2} + \rho_0 e_0\right) + p_0\, v_{0n}. \tag{11.28}$$

Let us assume the unperturbed field $(\rho_0, \mathbf{v}_0, p_0)$, s, and $\mathbf{n}$ are given. Our goal is to determine the perturbed field $(\rho, \mathbf{v}, p)$ as a solution to (11.26)–(11.28). To do so, we first introduce the *Mach number* instead of the shock velocity and the specific volume instead of ρ:

$$M_0 = \frac{s - v_{0n}}{c_0}, \qquad V = \frac{1}{\rho}, \tag{11.29}$$

where c_0 represents the unperturbed sound velocity:

$$c_0 = \sqrt{\gamma \frac{p_0}{\rho_0}}. \tag{11.30}$$

From Eq. (11.26), we obtain

$$V = V_0 \left(1 + \frac{v_{0n} - v_n}{c_0 M_0}\right), \tag{11.31}$$

and by substituting into Eq. (11.27), we have

$$\mathbf{v} = \mathbf{v}_0 + \frac{p - p_0}{c_0 M_0} V_0 \, \mathbf{n}. \tag{11.32}$$

By taking the scalar product of (11.32) with $\mathbf{n}$ and squaring both sides of (11.32), we obtain

$$v_n = v_{0n} + \tfrac{p-p_0}{c_0 M_0} V_0 \tag{11.33}$$

$$v^2 = v_0^2 + \tfrac{(p-p_0)^2}{c_0^2 M_0^2} V_0^2 + 2 \tfrac{p-p_0}{c_0 M_0} V_0 \, v_{0n}. \tag{11.34}$$

Substituting (11.33) into (11.31) and considering (11.30), we obtain

$$V = V_0 \left(1 - \frac{p - p_0}{M_0^2 p_0 \gamma}\right). \tag{11.35}$$

Note that so far we have expressed $\mathbf{v}$ and V in terms of p and M_0 (see (11.32) and (11.35)).

By inserting s from $(11.29)_1$, v_n and v^2 from (11.33) and (11.34), e and e_0 from $(8.36)_1$, c_0 from (11.30), and finally taking into account $(11.29)_2$ and (11.35), after some laborious but straightforward calculations, Eq. (11.28) becomes

$$\frac{(p - p_0) V V_0 \left\{(p - p_0)(1 + \gamma) - 2\gamma p_0 (M_0^2 - 1)\right\}}{2 M_0 (\gamma - 1)\sqrt{\gamma p_0 V_0}} = 0,$$

from which, in addition to the trivial solution $p = p_0$, we have the solution that gives the pressure after the shock as a function of the Mach number (see Fig. 11.7a):

$$p = p_0 \left(1 + \frac{2\gamma}{\gamma + 1} \left(M_0^2 - 1 \right) \right). \tag{11.36}$$

By substituting (11.36) into (11.35) and (11.32), we finally have

$$V = V_0 \left(1 - \frac{2}{\gamma + 1} \frac{M_0^2 - 1}{M_0^2} \right), \tag{11.37}$$

$$\mathbf{v} = \mathbf{v}_0 + \frac{2c_0}{\gamma + 1} \frac{M_0^2 - 1}{M_0} \mathbf{n} \tag{11.38}$$

(see Fig. 11.7b and c).

Equations (11.36)–(11.38) provide the perturbed field in terms of the unperturbed field and the Mach number. From (8.34), we can also obtain the perturbed temperature ϑ in terms of the unperturbed temperature (see Fig. 11.7d):

$$\vartheta = \vartheta_0 \left\{ 1 + 2\frac{(M_0^2 - 1)(\gamma M_0^2 + 1)(\gamma - 1)}{M_0^2 (1 + \gamma)^2} \right\}. \tag{11.39}$$

From (11.36) and (11.37), we obtain

$$\frac{[p]}{[V]} = -\frac{c_0^2 M_0^2}{V_0^2} < 0. \tag{11.40}$$

Therefore, there are two possibilities:

(i) $[p] > 0$ and $[V] < 0$: This case corresponds to $M_0^2 > 1$.
(ii) $[p] < 0$ and $[V] > 0$: This case corresponds to $M_0^2 < 1$.

Mathematically, both situations are acceptable, but which of the two is physically acceptable?

11.2.1 Entropy Growth and Admissible Shocks

In order to answer this question, let us first observe that for regular processes, every solution of the Euler system (8.5) is also a solution of the entropy balance equation:

$$\frac{\partial \rho S}{\partial t} + \frac{\partial (\rho S v_i)}{\partial x_i} = \frac{r}{\vartheta} \tag{11.41}$$

as long as (8.26) is satisfied. This means that we can interchange the energy balance laws $(8.5)_3$ with the entropy balance law (11.41) without changing the solutions. In other words, the solutions of (8.5) are the same as the solutions of the system formed by $(8.5)_{1,2}$ and (11.41).

But what happens if instead of having classical solutions, we have weak solutions, particularly shock waves? We can immediately verify that weak solutions are different!

In the case of shock waves, if the weak solutions of the system (8.5) are the same as the system $(8.5)_{1,2}$ plus (11.41), then the Rankine–Hugoniot equations should coincide, and therefore, the solutions (11.36)–(11.38) of (11.23)–(11.25) should also satisfy the Rankine–Hugoniot equation associated with (11.41), which is

$$\eta = s[\rho S] - [\rho S v_n] = [\rho(s - v_n)S] \tag{11.42}$$

and should be identically zero. However, η is not zero. In fact, from (11.26) and $(11.29)_1$, we have

$$\eta = \rho_0(s - v_{0n})[S] = \rho_0 c_0 M_0 [S],$$

and by inserting (11.36) and (11.37) into $(8.36)_2$, we obtain

$$\eta = \rho_0 c_0 c_V \, M_0 \log\left\{\left(\frac{2 + M_0^2(\gamma - 1)}{M_0^2(\gamma + 1)}\right)^{\gamma} \frac{2M_0^2\gamma + 1 - \gamma}{1 + \gamma}\right\}. \tag{11.43}$$

Figure 11.6 provides η as a function of M_0. Therefore, we have an important result that *if we modify the balance laws in an equivalent form for classical solutions, weak solutions—in general—are not the same!*

Since η represents the entropy production due to the shock , it is natural to require, in the spirit of the entropy principle, that the physically meaningful solutions of the shock wave type are only those for which

$$\eta > 0,$$

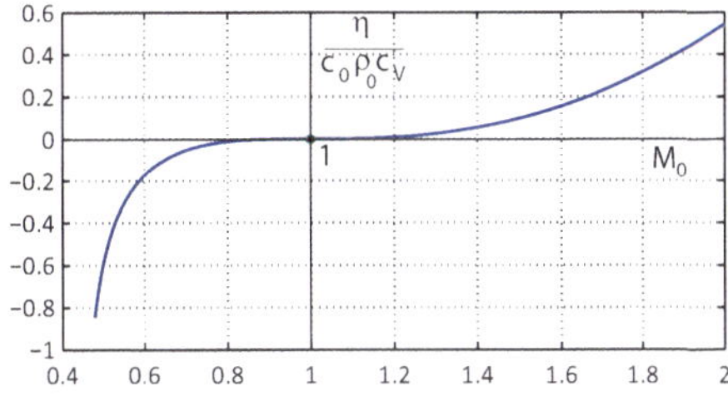

Fig. 11.6 Growth of entropy across the shock for $\gamma = 5/3$

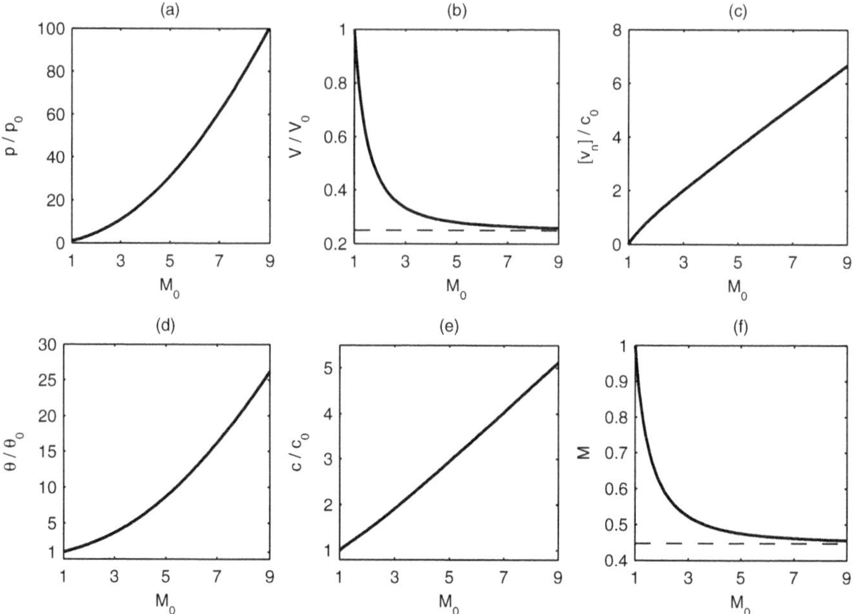

Fig. 11.7 Perturbed fields after the passage of the shock wave as functions of the Mach number

and thus, from (11.43), we necessarily have

$$M_0^2 > 1.$$

We will return later to the question of the different roles of the entropy principle in the case of classical solutions and in the case of weak solutions in Sect. 11.2.1. For now, we accept the shock wave only if $M_0^2 > 1$, and therefore, case (i) is the only physically admissible one: The pressure increases after the passage of the shock wave, while the volume decreases (*compression wave*). It is interesting to note that as $M_0 \to \pm\infty$, the volume ratio tends to a finite value:

$$\lim_{M_0 \to \pm\infty} \frac{V}{V_0} = \frac{\gamma - 1}{\gamma + 1}.$$

In the case of a monatomic gas for which $\gamma = 5/3$, this limit is $1/4$. In Fig. 11.7, the perturbed fields are represented as a function of the Mach number. Specifically, (a) shows the ratio of pressures after and before the shock, (b) represents the ratio of specific volumes, (c) displays the jump in normal velocity in terms of unperturbed sound velocity, (d) illustrates the ratio of temperatures, and (e) demonstrates the ratio of sound velocities.

Physically possible shocks are those with $M_0^2 > 1$ (supersonic shocks), one traveling in the direction of the x-axis with $M_0 > 1$ (shock S_5), and one going in the opposite direction with $M_0 < -1$ (shock S_1). Taking into account that the perturbed sound velocity is given by

$$c^2 = \gamma \frac{p}{\rho} = \gamma \frac{k}{m} \vartheta,$$

we obtain

$$\frac{c^2}{c_0^2} = \frac{\vartheta}{\vartheta_0}, \tag{11.44}$$

and therefore, we can deduce the perturbed Mach number as

$$M = \frac{s - v_n}{c}.$$

By considering (11.33), $(11.29)_1$, (11.39), and (11.44), we obtain the explicit expression for M^2 in terms of M_0:

$$M^2 = \frac{2 + M_0^2(\gamma - 1)}{1 + \gamma(2M_0^2 - 1)}$$

as a function of M_0. Since $M_0^2 > 1$, an interesting result is that

$$M^2 < 1.$$

Hence, shocks are supersonic on one side of the wavefront and subsonic on the other side. Finally, it should be noted that

$$\lim_{M_0 \to \infty} M = \sqrt{\frac{\gamma - 1}{2\gamma}},$$

and for a monatomic gas, the limit is $\sqrt{1/5}$. Figure 11.7f provides M as a function of M_0.

11.2.2 Characteristic Shocks

Examining the R-H Eqs. (11.26)–(11.28), we see that they also admit another solution:

$$v_n = v_{0n} = s, \quad p = p_0, \tag{11.45}$$

with

$$[\mathbf{v}_T] \text{ arbitrary}, \quad [\rho] \text{ arbitrary}, \quad \mathbf{v}_T = \mathbf{v} - v_n \mathbf{n}, \tag{11.46}$$

where $\mathbf{v}_T$ indicates the tangential component of velocity. Therefore, there exists a family of shocks that travel with the normal velocity of the fluid particles and depend on *three arbitrary parameters* (shocks $S_{2,3,4}$). These shocks are called *material or contact shocks*. Ultimately, we can represent the *characteristic shock* we have identified as follows:

$$p = p_0, \quad V = V_0 + \alpha, \quad \mathbf{v} = \mathbf{v}_0 + \mathbf{w}; \quad (\mathbf{w} \cdot \mathbf{n} = 0) \tag{11.47}$$

with $\alpha, \mathbf{w}$ as arbitrary parameters. It can be observed from (11.42) that for this characteristic shock, $\eta = 0$. *From a thermodynamic point of view, the characteristic shock corresponds to a reversible process: There is no entropy production through the shock.*

In general, we can define a characteristic shock as follows:

Definition 11.1 (Characteristic Shock) A shock is called characteristic if there exists an eigenvalue λ such that $s = \lambda(\mathbf{u}_0) = \lambda(\mathbf{u})$ with $\mathbf{u} \neq \mathbf{u}_0$ being a solution of the RANKINE–HUGONIOT equations.

It can be shown (BOILLAT [21]) that if an eigenvalue λ is exceptional (10.5), there always exists a characteristic shock with velocity $s = \lambda = \lambda_0$. It should be noted that in this special case, the shock velocity s is fixed and coincides with the characteristic velocity. Therefore, the characteristic shock corresponds to a vertical bifurcation of the trivial solution of zero shock (see Fig. 11.5).

The perturbed field depends on m parameters, where m is the multiplicity of the characteristic velocity. In the fluid example, $m = 3$ and the 3 parameters are α and $\mathbf{w}$.

11.3 Riemann Problem and Nonuniqueness of Weak Solutions

For conservation laws systems (11.1) with $\mathbf{f} \equiv 0$ and one spatial dimension, a solution of (11.1) corresponding to an initial data of the form

$$\mathbf{u}(x, 0) = \begin{cases} \mathbf{u}_0 \text{ for } x > 0 \\ \\ \mathbf{u}_1 \text{ for } x < 0, \end{cases} \tag{11.48}$$

with $\mathbf{u}_0$ and $\mathbf{u}_1$ constant vectors, is called a RIEMANN PROBLEM. As an example, consider the BURGERS' equation (10.53) rewritten in conservative form:

$$u_t + \left(\frac{u^2}{2}\right)_x = 0, \tag{11.49}$$

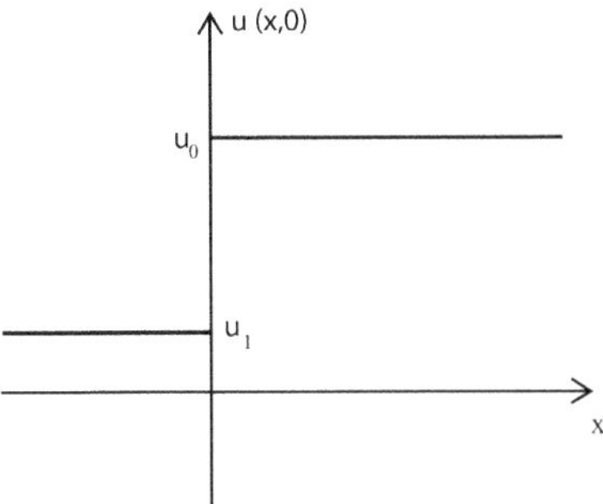

Fig. 11.8 Discontinuous initial data

and let us look for a solution with an initial Riemann data (shock initial condition) (Fig. 11.8).

If the sought solution is a shock, along the line $dx/dt = s$, the Rankine–Hugoniot equations must hold:

$$-s[u] + \left[u^2/2\right] = 0, \tag{11.50}$$

and therefore:

$$s = \frac{\left[u^2/2\right]}{[u]} = \frac{1}{2}(u_0 + u_1). \tag{11.51}$$

Therefore, there exists a weak solution represented by a shock wave (Fig. 11.9a):

$$u(x,t) = \begin{cases} u_0 & \text{for } x > st \\ u_1 & \text{for } x < st \end{cases} \quad \text{with } s = \frac{1}{2}(u_0 + u_1). \tag{11.52}$$

Now let us look for a solution u of (11.49) that depends on x and t through a single variable $z = x/t$. Requiring that $u \equiv u(z)$, we immediately have from (11.49)

$$u = z = x/t.$$

Therefore, there is another solution:

$$u(x,t) = \begin{cases} u_1 & \text{for } x < u_1 t \\ \frac{x}{t} & \text{for } u_1 t \le x \le u_0 t \\ u_0 & \text{for } x > u_0 t. \end{cases} \tag{11.53}$$

This solution satisfies the initial data, is continuous everywhere, and is differentiable (except for $t = 0$). Since this solution is a classical solution, it is also a weak solution and is called a *rarefaction wave* (see Fig. 11.9b). In conclusion, we have found two

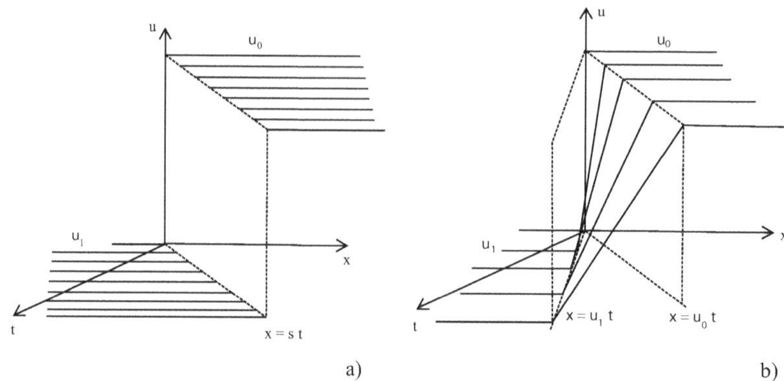

Fig. 11.9 Shock wave and rarefaction in space-time

weak solutions of (11.49) that satisfy the same Riemann-type initial conditions. However, in physical problems, only one solution is allowed.

It should be noted that the characteristic curves are bundles of parallel lines with equations $x = u_0 t + x_0$ and $x = u_1 t + x_0$. However, there are two possibilities:

(a) If $u_1 < u_0$, the two characteristic families do not intersect in the future (Fig. 11.10a).
(b) If $u_1 > u_0$, the two characteristic families intersect (Fig. 11.10b).

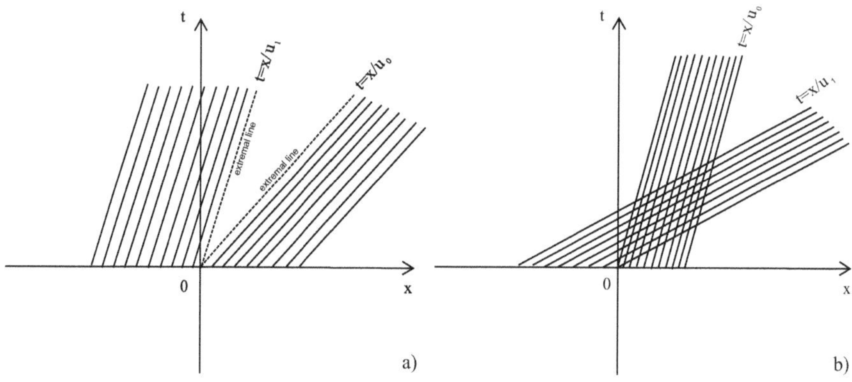

Fig. 11.10 Rarefaction or shock? Lax conditions

11.4 Lax Conditions and Entropy Growth

The Lax conditions [22] provide a criterion for selecting between a rarefaction wave and a shock wave. Considering the case of a scalar equation, the situation depicted in Fig. 11.10a corresponds to the condition:

$$\lambda(u_1) < \lambda(u_0), \tag{11.54}$$

while the one in Fig. 11.10b corresponds to

$$\lambda(u_1) > \lambda(u_0). \tag{11.55}$$

LAX deduced that if condition (11.54) is satisfied, then the physically acceptable solution for the Riemann problem is the rarefaction wave. On the other hand, if condition (11.55) is satisfied, then the physically acceptable solution is the shock wave. Therefore, for an initial shock wave to propagate (*stable shock*), condition (11.55) must hold. In the case of the Burgers' equation where $\lambda(u) = u$, if $u_1 > u_0$, then the physical solution is the shock wave (11.52), and if $u_1 < u_0$, then the solution is the rarefaction wave (11.53).

To understand the physical meaning of the conditions (11.55) for a stable shock wave, let us revisit the discussion on entropy growth through a shock wave (see Sect. 11.2.1). Considering again the example of the Burgers' equation, we multiply (11.49) by u to obtain an alternative form (among infinitely many) of the conservation law (11.49):

$$\left(\frac{u^2}{2}\right)_t + \left(\frac{u^3}{3}\right)_x = 0. \tag{11.56}$$

We can interpret (11.56) as the entropy equation associated with (11.49), and any smooth solution of (11.49) is also a solution of (11.56). However, when considering weak solutions and specifically shock waves, we have already seen in Sect. 11.2.1 that weak solutions are not unique. The solution of the Rankine–Hugoniot equations associated with (11.49) is given by (11.51) (replacing s as the parameter):

$$u_1 = 2s - u_0. \tag{11.57}$$

If this solution were also a solution of the Rankine–Hugoniot equations associated with (11.56), then the quantity

$$\eta = -s\left[\frac{u^2}{2}\right] + \left[\frac{u^3}{3}\right] \tag{11.58}$$

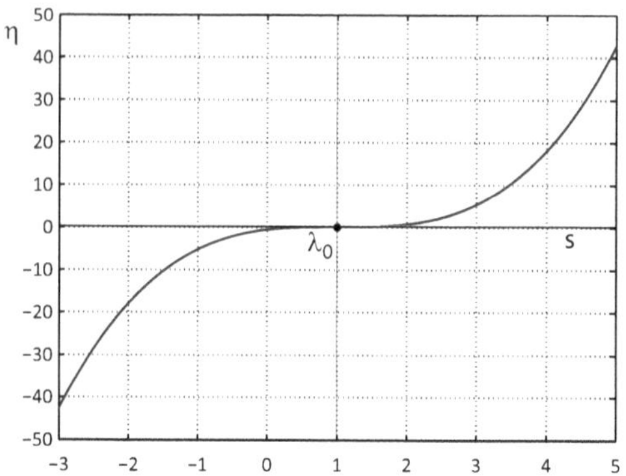

Fig. 11.11 Entropy production across a shock

should be identically zero. However, substituting (11.57) into (11.58), we obtain (see Fig. 11.11)

$$\eta = \frac{2}{3}(s - u_0)^3, \tag{11.59}$$

and thus $\eta \neq 0$.

What we have shown is that even though the phenomenon is reversible under differentiability conditions, there is entropy production through a shock wave because a shock wave is an irreversible process. The second law of thermodynamics must be satisfied through the shock wave, so the acceptable branch in Fig. 11.11 is the one for which $\eta > 0$ (*entropy growth criterion*).

In the discussed case, the admissible shocks are those for which $\eta > 0$ (i.e., $s > u_0$). However, if $s > u_0$, considering (11.51), it must hold that $s < u_1$. Therefore, stating that $\eta > 0$ is equivalent to saying that the admissible shocks are those for which:

$$u_0 < s < u_1. \tag{11.60}$$

Since in this example $\lambda = u$, condition (11.60) becomes

$$\lambda_0 < s < \lambda, \tag{11.61}$$

which are the *Lax conditions*.

Thus, through this specific example, we have verified that the Lax conditions are equivalent to *entropy growth* through the shock wave.

In the case of the Burgers' equation, the Lax conditions are also justified by the *artificial viscosity method*, which transforms the given problem into a parabolic problem by adding a second-order term of the form νu_{xx} to the Burgers' equation (11.68):

$$u_t + uu_x = \nu u_{xx}. \tag{11.62}$$

The solution now depends on the parameter ν, called the *artificial viscosity coefficient*. In the parabolic case (11.62), the weak solution is unique, and it can be shown that as $\nu \to 0$ with Riemann initial data (11.48), we obtain the solution of the hyperbolic problem obtained by applying the Lax criterion. The principle of entropy plays two important roles:

- For classical solutions, it selects acceptable constitutive equations.
- In the class of weak solutions, it selects admissible processes.

11.5 Traffic Flow

Continuum mechanics can also serve as a reference model for completely different issues in some situations. For example, the LIGHTHILL–WHITHAM–RICHARDS (LWR) model is a fundamental traffic flow model developed independently by LIGHTHILL and WHITHAM in 1955 [23] and by RICHARDS in 1956 [24]. It is one of the earliest and most widely used models in traffic flow theory. The model consists of considering the motion of cars as the flow of a fluid. Let us consider one spatial dimension and denote $\rho(x,t)$ the density of cars. Note that we are not keeping track of the individual cars, but just of the average number of cars per unit length of road. It is assumed that no vehicles enter or exit the road section under consideration, thus obeying the conservation principle of the number of cars $(8.5)_1$:

$$\frac{\partial \rho}{\partial t} + \frac{\partial \rho v}{\partial x} = 0. \tag{11.63}$$

In the simplest model, the velocity v is assumed to be a *constitutive equation* for ρ:

$$v \equiv v(\rho). \tag{11.64}$$

By inserting (11.64) into (11.63), we obtain a single differential equation for $\rho(x,t)$:

$$\frac{\partial \rho}{\partial t} + \frac{\partial a(\rho)}{\partial x} = 0; \quad \text{with} \quad a(\rho) = \rho v(\rho). \tag{11.65}$$

Equation (11.65) is a hyperbolic equation with a characteristic velocity

$$\lambda(\rho) = a'(\rho) = v(\rho) + \rho\, v'(\rho). \tag{11.66}$$

The typical law assumed for (11.64) is a linear equation:

$$v(\rho) = v_m\left(1 - \frac{\rho}{\rho_m}\right), \tag{11.67}$$

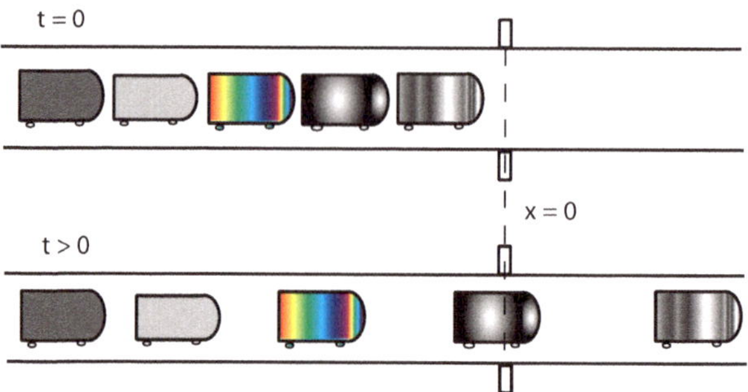

Fig. 11.12 Motion of cars when the traffic light turns green

where ρ_m is the maximum density corresponding to closely packed and stationary vehicles, $v(\rho_m) = 0$, while for $\rho = 0$ (free road), $v(0) = v_m$ represents the maximum velocity. In the case where (11.67) holds, we have

$$a(\rho) = \rho\, v_m \left(1 - \frac{\rho}{\rho_m}\right), \qquad \lambda(\rho) = v_m \left(1 - 2\frac{\rho}{\rho_m}\right). \tag{11.68}$$

11.5.1 The Traffic Light Problem

One of the most interesting problems is finding a solution to (11.65) whose initial data corresponds to a group of cars stopped at a red traffic light, which turns green immediately after $t = 0$:

$$\rho(x, 0) = \rho_0(x) = \begin{cases} \rho_m & \text{for } x < 0 \\ 0 & \text{for } x > 0. \end{cases} \tag{11.69}$$

This is a Riemann problem (see Fig. 11.12). Since from (11.68) we have $\lambda(0) = v_m$ and $\lambda(\rho_m) = -v_m$, the conditions (11.54) are satisfied, and therefore, the characteristics are as shown in Fig. 11.10a, and the Lax conditions are violated. As mentioned in the previous paragraph, the shock wave is unstable and decays into a rarefaction wave. By seeking solutions of (11.65) that depend on $z = x/t$, we have in this case

$$\frac{x}{t} = a'(\rho) = \lambda(\rho) \implies \rho = \frac{\rho_m}{2}\left(1 - \frac{x}{v_m t}\right),$$

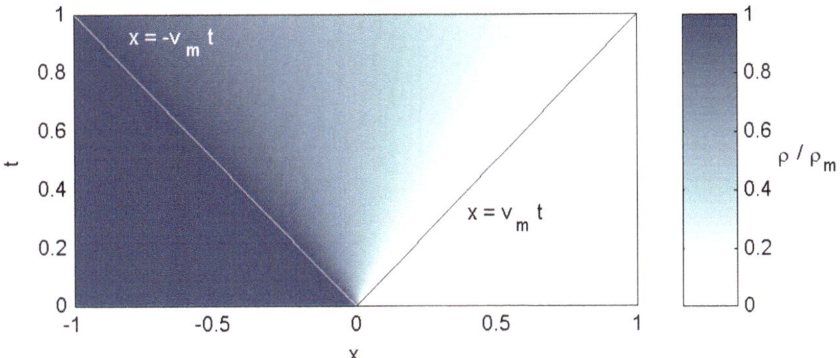

Fig. 11.13 Rarefaction wave in space-time

and therefore, the only weak solution of the Riemann problem for case (11.67) is

$$\rho(x,t) = \begin{cases} \rho_m & \text{for } x < -v_m t \\ \widetilde{\rho} = \frac{\rho_m}{2}\left(1 - \frac{x}{v_m t}\right) & \text{for } -v_m t \le x \le v_m t\,. \\ 0 & \text{for } x > v_m t \end{cases} \tag{11.70}$$

Immediately after $t = 0$, the shock disappears, and the cars start slowly, following the profile of the rarefaction wave, as shown in Fig. 11.12.

In space-time, the solution is represented in Fig. 11.13. The grayscale represents the ratio between the density and the maximum density, ranging from white (corresponding to a value of 0) to black (corresponding to 1).

A more realistic problem arises when considering that only the cars stopped at the traffic light are within a finite interval $x \in [-l, 0]$. In this case, we have a double Riemann problem with initial jumps at $x = -l$ and $x = 0$:

$$\rho(x,0) = \rho_0(x) = \begin{cases} \rho_m & \text{for } -l < x < 0 \\ 0 & \text{otherwise} \end{cases}.$$

The initial profile is a square wave represented in Fig. 11.14. We have seen that the initial shock at $x = 0$ is unstable, while the shock at $x = -l$ is stable by reversing the conditions of Lax:

$$\lambda(0) > \lambda(\rho_m).$$

The Rankine–Hugoniot equation for (11.65) is

$$s = \frac{a(\rho_1) - a(\rho_0)}{\rho_1 - \rho_0}.$$

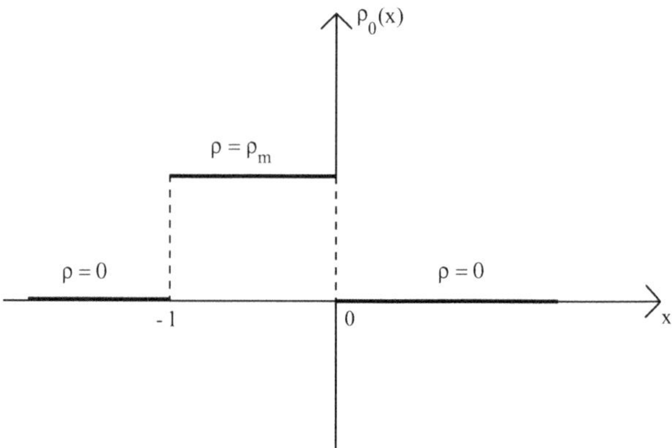

Fig. 11.14 Initial profile

It follows that until the time t^* (where the characteristic $x = -v_m t$ intersects $x = -l$), we have

$$t^* = \frac{l}{v_m},$$

and:

$$s = \frac{a(0) - a(\rho_m)}{-\rho_m} = 0.$$

Therefore, the solution until time t^* is given by (see Fig. 11.15)

$$\rho(x,t) = \begin{cases} 0 & \text{for} \quad x < -l \\ \rho_m & \text{for} \quad -l < x < -v_m t \\ \widetilde{\rho} = \frac{\rho_m}{2}\left(1 - \frac{x}{v_m t}\right) & \text{for} \quad -v_m t \le x \le v_m t \\ 0 & \text{for} \quad x > v_m t. \end{cases} \tag{11.71}$$

In space-time, the situation is described by Fig. 11.16.

For $t > t^*$, to the left of the shock, $\rho = 0$, while to the right, there will be a rarefaction $\rho = \widetilde{\rho} = \frac{\rho_m}{2}\left(1 - \frac{x}{v_m t}\right)$. Thus, for $t > t^*$:

$$s = \frac{a(0) - a(\widetilde{\rho})}{0 - \widetilde{\rho}} = v(\widetilde{\rho}) = \frac{v_m}{2}\left(1 + \frac{x}{t v_m}\right). \tag{11.72}$$

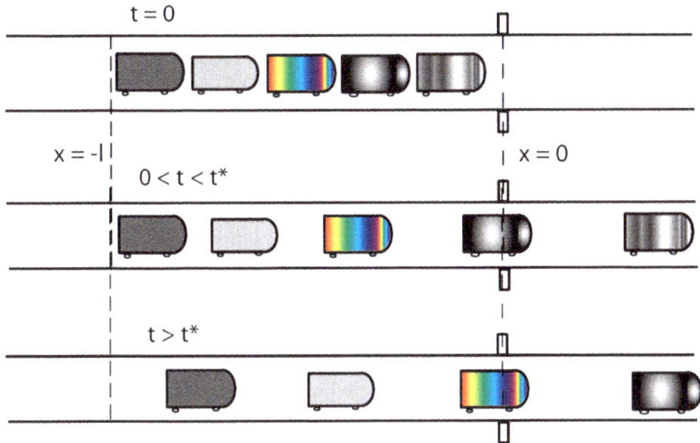

Fig. 11.15 Car motion after light becomes green

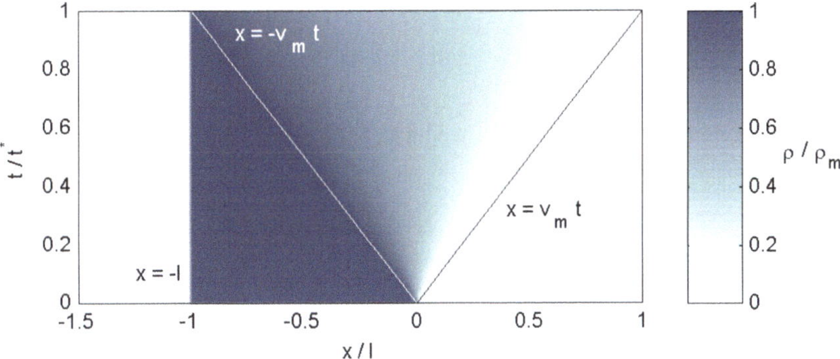

Fig. 11.16 Solution in the space-time before t^*

The shock line is no longer the line $x = -l$, but a solution of:

$$\frac{dx}{dt} = s = \frac{v_m}{2}\left(1 + \frac{x}{t v_m}\right); \quad x(t^*) = -l. \tag{11.73}$$

By making the change of variables $y = x - v_m t$, Eq. (11.73) becomes

$$\frac{dy}{dt} = \frac{y}{2t}; \quad y(t^*) = -2l, \tag{11.74}$$

which has the solution:

$$y = -2\sqrt{l v_m t},$$

and therefore:

$$x = x^*(t) = v_m t - 2\sqrt{l v_m t} \tag{11.75}$$

(see Figs. 11.17 and 11.15). The solution for $t \leq t^*$ is given by (11.71), while for $t > t^*$, it is

$$\rho(x,t) = \begin{cases} \widetilde{\rho} = \frac{\rho_m}{2}\left(1 - \frac{x}{v_m t}\right) & \text{for } x^*(t) < x < v_m t \\ \\ 0 & \text{otherwise} \end{cases} \tag{11.76}$$

with $x^*(t)$ given by (11.75). The solution in space-time is represented in Fig. 11.17. It can be observed that the last car starts moving at time t^* and from (11.75) reaches the traffic light $x = 0$ at time $\widetilde{t}$, where

$$t^* = \frac{l}{v_m} \qquad \text{and} \qquad \widetilde{t} = \frac{4l}{v_m} = 4t^*.$$

Substituting (11.75) into (11.72), we have

$$s(t) = v_m\left(1 - \sqrt{\frac{l}{v_m t}}\right) = v_m\left(1 - \sqrt{\frac{t^*}{t}}\right). \tag{11.77}$$

The (11.77) is in fact the velocity of last car, and the graph is reported in the left part of Fig. 11.18, while the acceleration

$$\dot{s} = \frac{1}{2}\sqrt{\frac{l v_m}{t^3}} = \frac{1}{2}\frac{v_m}{t^*}\left(\frac{t^*}{t}\right)^{\frac{3}{2}} \tag{11.78}$$

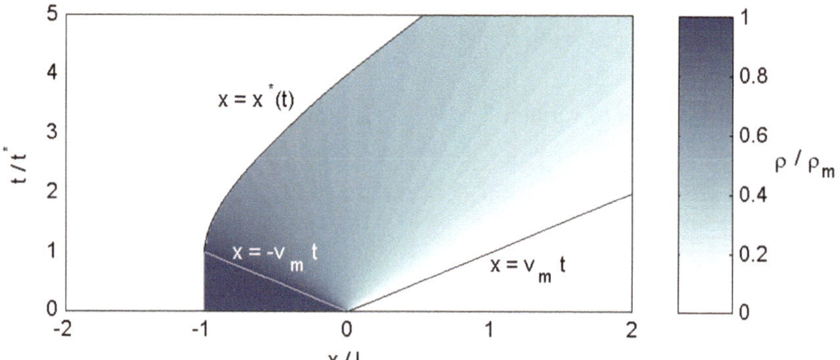

Fig. 11.17 Solution in the space-time

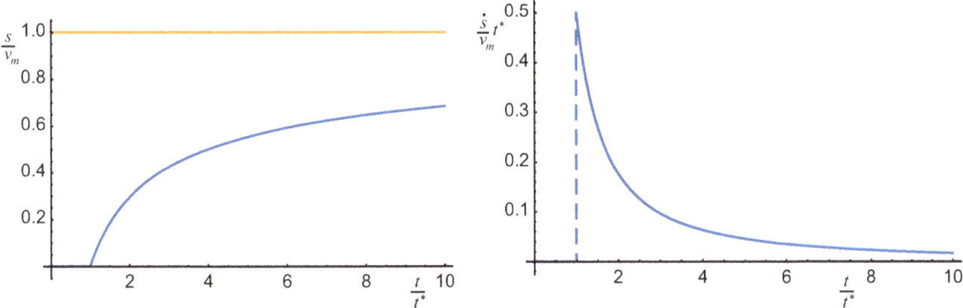

Fig. 11.18 Dimensionless velocity (left) and dimensionless acceleration (right) of last car

is reported in the right side: the acceleration jump at $t = t^*$ from 0 to $v_m^2/(2l)$ and then decay.

The jump in ρ is given by (S is the shock line defined by Eq. (11.75))

$$[\rho] = 0 - \tilde{\rho}\Big|_{S^+} = -\rho_m \sqrt{\frac{l}{v_m t}}. \tag{11.79}$$

From Eqs. (11.77) and (11.79), it can be seen that the shock after t^* moves at the speed of the last car (see Figs. 11.15, 11.18 and 11.17), and for $t \to \infty$, it tends to the maximum velocity v_m. Meanwhile, the shock amplitude decreases and tends to become zero. For long times, all cars tend to travel at the velocity v_m, and the density jump tends to zero.

The initial profile given in Fig. 11.14 as time increases is represented in Fig. 11.19. For large time values, this result is in agreement with a general theorem by Liu [25], which guarantees (in the case under consideration) that the initial square wave converges to zero, forming a triangle whose base tends to infinity and height tends to zero in such a way that the area $A = \rho_m l$ remains constant for any $t \geq 0$. In fact from Fig. 11.19, we have the following coordinates:

in the first line, $A \equiv (-l, 0)$, $B \equiv (-l, \rho_m)$, $C \equiv (0, \rho_m)$, $D \equiv (0, 0)$,

in the second line, $A \equiv (-l, 0)$, $B \equiv (-l, \rho_m)$, $C \equiv (-v_m t, \rho_m)$, $D \equiv (v_m t, 0)$,

in the third line, $A \equiv (-l, 0)$, $B \equiv (-l, \rho_m)$, $D \equiv (l, 0)$,

in the last line, $A \equiv (x^*(t), 0)$, $B \equiv (x^*(t), \tilde{\rho}(x^*(t), t))$, $D \equiv (v_m t, 0)$,

and taking into account of the expression of $x^*(t)$ given in (11.75) and $\tilde{\rho}$ given in (11.76) is easy to verify that in all figures the area is always $A = \rho_m\, l$.

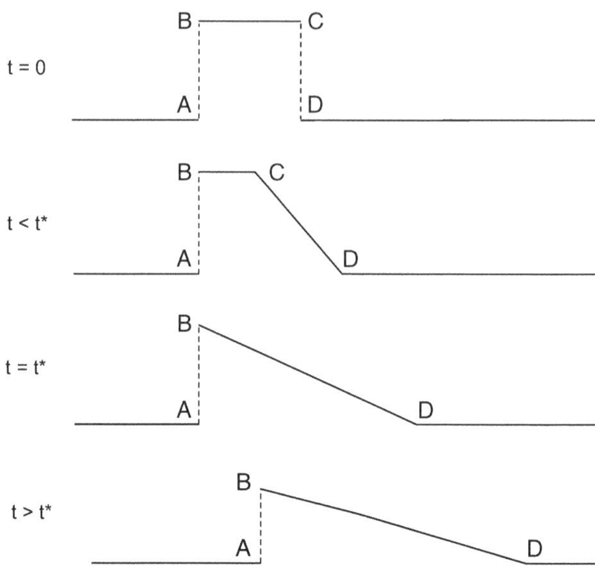

Fig. 11.19 Sketch of the change in density for increasing time

11.6 Riemann Problem for a General System

In the previous examples of Riemann problems, only the case of a scalar equation was considered, and the Lax criterion determined whether the initial shock could propagate or transform into a rarefaction. It is interesting to see what happens when the one-dimensional system of conservation laws consists of N equations, i.e.,

$$\mathbf{u}_t + \partial_x \mathbf{F}(\mathbf{u}) = 0,$$

with an initial condition:

$$\mathbf{u}(x, 0) = \begin{cases} \mathbf{u}_0 & \text{for } x > 0 \\ \mathbf{u}_1 & \text{for } x < 0 \end{cases}$$

(where $\mathbf{u}_0$ and $\mathbf{u}_1$ are constant vectors). In this case, it can be shown that the initial shock *splits* into a combination of shocks and rarefactions (connected by constant states) depending on whether the individual waves satisfy the Lax conditions or not [22].

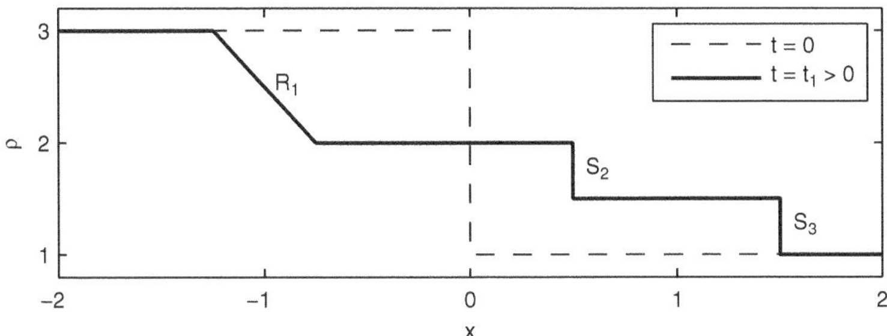

Fig. 11.20 Sketch of the mass density solution of the Riemann problem for Euler system

For example, in the case of a one-dimensional Euler fluid, where $\lambda_0^{(1)} = v_0 - c_0$, $\lambda_0^{(2)} = v_0$, $\lambda_0^{(3)} = v_0 + c_0$, the initial jump will typically split into a sonic shock S_3 and a contact shock wave S_2 propagating in one direction, and a rarefaction R_1 propagating in the other direction, as qualitatively indicated in Fig. 11.20.

Chapter 12
Beyond Classical Thermomechanics

Abstract We present other problematics on continuum mechanics that are not considered in the previous chapters.

12.1 Outlook

The previous chapters have focused on classical thermomechanics of *simple materials*. However, it is worth mentioning that the mathematical modeling of materials sometimes requires going beyond classical treatments. This is the case, for example, with *plastic* materials, *viscoelastic* materials, *granular* materials, and materials with *memory*.

Furthermore, it is often necessary to consider that particles have nonzero dimensions. This gives rise to a whole field of study that has as its prototype the COSSERAT *continuum (1909)* [26], in which a particle is envisioned as a very small rigid body that also possesses spin. To maintain a continuous distribution of material points, an associated tetrahedron can be assigned to each point, which figuratively represents a frame attached to the rigid body associated with the particle. This allows us to consider not only the displacement field but also the rotation (expressed, for example, through Euler angles) experienced by the tetrahedron in the transition from C^* to C (see Fig. 12.1).

Such continua are therefore called *structured continua* or *polar continua*. In this case, the momentum equation becomes a true field equation and no longer implies that the stress tensor is symmetric. For this reason, the theory is also known as the *asymmetric theory* (GRIOLI-1960) [27] .

The Cosserat theory has further generalizations such as the *Microstructure theory* (see, e.g., [28, 29]) or specializations such as the *Liquid Crystal theory* (see, for example, [30, 31]).

Regarding fluids, there is a wide class of fluids known as *non-Newtonian fluids*, where it is not legitimate to assume a linear relationship between the viscous stress tensor and the velocity gradient (see, for example, [32]).

© The Author(s), under exclusive license to Springer Nature Switzerland AG 2024

T. Ruggeri, *Introduction to the Thermomechanics of Continua and Hyperbolic Systems*, La Matematica per il 3+2 167,
https://doi.org/10.1007/978-3-031-69951-1_12

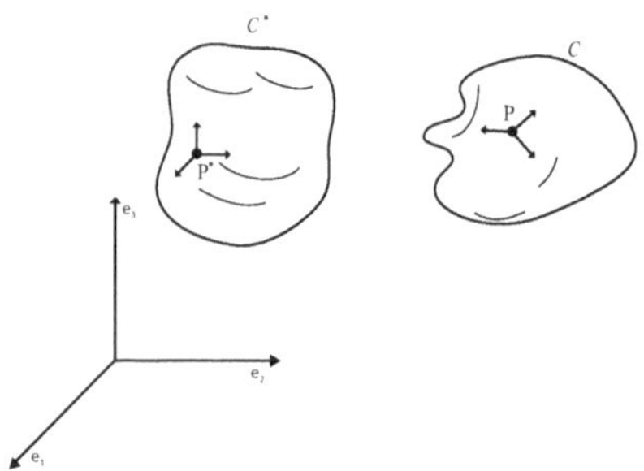

Fig. 12.1 Polar continuum

Finally, it should be noted that in the case of rarefied gases, kinetic theory is commonly used in addition to the continuum approach. The celebrated *Boltzmann equation* describes the evolution of the distribution function $f(\mathbf{x}, \mathbf{c}, t)$, which represents the probability of finding a particle with a microscopic velocity $\mathbf{c}$ at position $\mathbf{x}$ and time t. The Boltzmann equation has the following form:

$$\frac{\partial f}{\partial t} + c_i \frac{\partial f}{\partial x_i} + b_i \frac{\partial f}{\partial c_i} = Q, \tag{12.1}$$

where b_i represents external forces, and Q is the *collision term*, which is an integral extremely complex and takes into account the interactions between particles.

Macroscopic quantities are obtained as *moments* of the distribution function. For example, the material density is obtained by integrating f over velocity space:

$$\rho(\mathbf{x}, t) = \int_{\mathbb{R}^3} f(\mathbf{x}, \mathbf{c}, t)\, d\mathbf{c}.$$

Similarly, the momentum $\rho\mathbf{v}$ is given by

$$\rho v_i = \int_{\mathbb{R}^3} f(\mathbf{x}, \mathbf{c}, t) c_i\, d\mathbf{c},$$

and other macroscopic quantities can be defined in a similar way.

Since f is a solution of the Boltzmann equation (12.1), the moments satisfy partial differential equations that involve both $\mathbf{x}$ and t. It is interesting to note that, due to the special properties of Q, considering the first five moments exactly reproduces the balance laws of mass, momentum, and energy in continuum thermomechanics. This proves the existence of an intimate connection between continuum mechanics and kinetic theory.

Efforts have been made to find the points of contact between the macroscopic approach of continuum mechanics and the kinetic approach, especially when truncating the moments beyond the first five equations. The correlation between the two theories requires a complex analysis that goes beyond the scope of this book and forms the basis of the so-called *Rational Extended Thermodynamics* [19, 20].

It is worth mentioning that the equations of Fourier and Navier–Stokes in Rational Extended Thermodynamics are no longer constitutive equations but approximations (when certain relaxation times are negligible) of true evolution laws. To get an idea of these results, consider the Cattaneo equation (9.12), which *approximates* the Fourier equation (9.2) when the relaxation time τ is negligible.

References

1. Biscari, P., Ruggeri, T., Saccomandi, G., Vianello, M.: Meccanica Razionale, 4 edizione. Springer, Berlin (2022). An English version of this book is currently being prepared
2. Truesdell, C., Toupin, R.A.: The classical field theories. In: Handbuch der Physik, vol. 3/l. Springer, Berlin (1960)
3. Liu, I.-S.: Continuum Mechanics. Springer, Berlin (2010)
4. Gurtin, M.E., Fried, E., Anand, L.: The Mechanics and Thermodynamics of Continua. Cambridge University, Cambridge (2013)
5. Holzapfel, G.A.: Nonlinear Solid Mechanics: A Continuum Approach for Engineering. Wiley, New York (2000)
6. Temam, R., Miranville, A.: Mathematical Modeling in Continuum Mechanics. Cambridge University, Cambridge (2005)
7. Manacorda, T.: Introduzione alla Meccanica dei Continui. Pitagora, Bologna (1979)
8. Banfi, C.: Introduzione alla meccanica dei continui. Cedam, Padova (1990)
9. Ciarletta, M., Iesan, D.: Elementi di meccanica dei continui con applicazioni. Pitagora (1997)
10. Forte, S., Preziosi, L., Vianello, M.: Meccanica dei Continui. Springer, Berlin (2019)
11. Ruggeri, T.: Galilean invariance and entropy principle for systems of balance laws. Continuum Mech. Thermodyn. **1**, 3 (1989)
12. Coleman, B.D., Noll, W.: The thermodynamics of elastic materials with heat conduction and viscosity. Arch. Ration. Mach. Analysis **13**, 167 (1963)
13. Müller, I.: On the entropy inequality. Arch. Rational Mech. Anal. **26**, 118 (1967)
14. Müller, I.: The History of Thermodynamics. Springer, Berlin (2007)
15. Mooney, M.: A theory of large elastic deformation. J. Appl. Phys. **11**(9), 582 (1940)
16. Rivlin, R.S.: Large elastic deformations of isotropic materials. IV. Further developments of the general theory. Philos. Trans. R. Soc. London, Ser. A: Math. Phys. Sci. **241**(835), 379 (1948)
17. Ogden, R.W.: Large deformation isotropic elasticity—on the correlation of theory and experiment for incompressible rubberlike solids. Proc. R. Soc. Lond. A **326**, 565 (1972)
18. Müller, I.: Thermodynamics, Pitman, London (1985)
19. Müller, I., Ruggeri, T.: Rational extended thermodynamics. In: Springer Tracts in Natural Philosophy, vol. 37, 2nd edn., p. 383. Springer, Berlin (1998)
20. Ruggeri, T., Sugiyama, M.: Classical and Relativistic Rational Extended Thermodynamics. Springer, Berlin (2021)
21. Boillat, G.: Chocs caractéristiques. C. R. Acad Sci. Paris A **274**, 1018 (1972)

© The Author(s), under exclusive license to Springer Nature Switzerland AG 2024

T. Ruggeri, *Introduction to the Thermomechanics of Continua and Hyperbolic Systems*, La Matematica per il 3+2 167,
https://doi.org/10.1007/978-3-031-69951-1

22. Lax, P.D.: Hyperbolic systems of conservation laws and the mathematical theory of shock waves CBMS-NSF. In: Regional Conference Series in Applied Mathematics, vol. 11. SIAM, New York (1973)
23. Lighthill, M.J., Whitham, G.B.: On kinematic waves ii. a theory of traffic flow on long crowded roads. Proc. R. Soc. Lond. A Math. Phys. Eng. Sci. **229**(1178), 317 (1955)
24. Richards, P.I.: Shock waves on the highway. Oper. Res. **4**, 42 (1956)
25. Liu, T.-P.: Linear and nonlinear large-time behavior of solutions of general systems of hyperbolic conservation laws. Commun. Pure Appl. Math. **30**, 767 (1977)
26. Cosserat, E., Cosserat, F.: Sur la théorie de l'élasticité. Premier mémoire. Annales de la faculté des sciences de Toulouse 1re série **10**(3-4), 1 (1896)
27. Grioli, G.: Elasticità Asimmetrica. Ann. Mat. Pura Appl. **50**, 389 (1960)
28. Eringen, A.C.: Microcontinuum Field Theories: I. Foundations and Solids. Springer, Berlin (1999)
29. Capriz, G.: Continua with Microstructure. In: Springer Tracts in Natural Philosophy, vol. 35 (2015)
30. de Gennes, P.G.: Short range order effects in the isotropic phase of nematic liquid crystals. Mol. Cryst. Liq. Cryst. **12**(1-2), 193 (1971)
31. Chandrasekhar, S.: Liquid Crystals. Cambridge University, Cambridge (1992)
32. Chhabra, R.P., Richardson, J.F.: Non-Newtonian Flow and Applied Rheology: Engineering Applications. Butterworth-Heinemann, Oxford (2008)

Author Index

© The Author(s), under exclusive license to Springer Nature Switzerland AG 2024

T. Ruggeri, *Introduction to the Thermomechanics of Continua and Hyperbolic Systems*, La Matematica per il 3+2 167,
https://doi.org/10.1007/978-3-031-69951-1

Analitical Index

© The Author(s), under exclusive license to Springer Nature Switzerland AG 2024

T. Ruggeri, *Introduction to the Thermomechanics of Continua and Hyperbolic Systems*, La Matematica per il 3+2 167,
https://doi.org/10.1007/978-3-031-69951-1

GPSR Compliance

The European Union's (EU) General Product Safety Regulation (GPSR) is a set of rules that requires consumer products to be safe and our obligations to ensure this.

If you have any concerns about our products, you can contact us on ProductSafety@springernature.com

In case Publisher is established outside the EU, the EU authorized representative is:

Springer Nature Customer Service Center GmbH
Europaplatz 3
69115 Heidelberg, Germany

Batch number: 10370708

Printed by Printforce, the Netherlands